主动式太阳能建筑营造

曲翠松 著

CUISONG QU

Active Building Construction

How the EnergyPLUS Home 4.0 was made

中国电力出版社
CHINA ELECTRIC POWER PRESS

内 容 提 要

同济大学联手德国的战略合作伙伴达姆施塔特工业大学共同参加了 2018 中国国际太阳能十项全能竞赛，经过 2 年半的设计及预制施工准备，成功打造了国内第一幢主动式太阳能建筑"正能量房 4.0"。本书详细介绍此项目的设计与建设过程，分为主动式·被动式、设计·理念、能量系统、材料、技术、预制、施工和实测八个章节，为主动式建筑在我国的后续实践提供参考。

本书用中英文撰写，面对与绿色建筑和建筑节能相关的中外广大建筑从业者，包括建筑设计人员、施工人员和材料供应单位；同时为建筑科研人员提供专项研究参考。建筑相关专业在校学生以及未来国际太阳能十项全能竞赛的参赛高校均可从中获益。

Summery

Tongji University and Technical University of Darmstadt participated as a joint team in the 2018 China International Solar Decathlon Competition. After 30 months of design and prefabrication, the team created the first active house in China, "EnergyPLUS Home 4.0". This book tells the story of the house's design and construction and aims to serve as a reference for future active buildings in China.

The book is written in Chinese and English. Potential readers include architects, builders, suppliers and students of architecture with an interest in energy-saving technology.

图书在版编目（CIP）数据

主动式太阳能建筑营造 / 曲翠松著. —北京：中国电力出版社，2020.1
ISBN 978-7-5198-4031-0

Ⅰ.①主… Ⅱ.①曲… Ⅲ.①太阳能建筑－建筑工程 Ⅳ.① TU18

中国版本图书馆 CIP 数据核字（2019）第 253241 号

出版发行：中国电力出版社
地　　址：北京市东城区北京站西街 19 号（邮政编码 100005）
网　　址：http://www.cepp.sgcc.com.cn
责任编辑：王　倩（010-63412607）
责任校对：黄　蓓　闫秀英
装帧设计：锋尚设计
责任印制：杨晓东
印　　刷：北京博海升彩色印刷有限公司
版　　次：2020 年 1 月第一版
印　　次：2020 年 1 月北京第一次印刷
开　　本：710 毫米 ×980 毫米　16 开本
印　　张：7.5
字　　数：171 千字
定　　价：65.00 元

前 言
Preface

在 2018 年中国国际太阳能十项全能竞赛中，同济大学联手德国的战略合作伙伴达姆施塔特工业大学共同参加，旨在通过主动式太阳能建筑的设计理念打造适合中国国情的高能效绿色建筑。参赛项目命名为"正能量房 4.0"，经过两年半的设计及预制施工准备，项目按照预期建成并参加比赛，其能量平衡与舒适度测试都达到了设计效果要求。基于主动式在我国还是比较新的概念，建成项目也很少，本书就这个项目的设计与建设过程进行详细分解介绍，为主动式建筑在我国的实践提供参考。

全书分 8 个章节。第 1 章介绍节能建筑的历史概况和发展方向，是本书主体部分的理论基础；从第 2 章起介绍项目的方方面面，包括设计理念、能量系统、材料、技术、预制、施工和实测 7 个主题。

设计理念部分在第 1 章介绍了达姆施塔特工业大学团队两次参加国际太阳能竞赛的冠军作品，第 2 章讲述正能量房 4.0 的空间理念，通过建筑空间设计尽量被动式利用太阳能，用"盒子"尝试整体式装配，并发展体现建筑风格，满足节能需求的外围护结构设计策略。与空间设计密切结合的是建筑的能量系统，在第 3 章中进行了介绍。其技术策略包含主动式和被动式两部分，是在分析建筑所在地的气候条件后得出的。第 4 章材料分 11 个小节，分别介绍建筑使用哪些主要材料实现其空间和能量概念。第 5 章技术一章阐述建筑所使用的技术措施和设备。第 6 章与第 7 章分别介绍建筑的预制过程和运往基地后的快速建造过程。第 8 章是建筑建成后在比赛期间实际测量记录的性能数据及其分析。

正能量房 4.0 所使用的材料和技术设备基本上都来自于国内的厂家，他们的产品和技术都是针对建筑节能需求研发和生产的，在各自的专业领域处于领先地位。书中提供了他们的信息，包括网址链接，希望对于想要深入了解的读者有所帮助。

基于主动式建筑与正能量房 4.0 的国际性，本书用中英双语撰写。考虑到中英专业读者对信息的不同需求，以及特殊环境语言的识别性，英文中略去少量对英文专业读者不甚重要的信息；部分使用英文软件做的技术分析图，未做中文翻译。

In the 2018 China International Solar Decathlon Competition, Tongji University and Technical University of Darmstadt participated as a joint team to create an energy-efficient green building using active solar technology. After two and a half years of design and prefabrication, the project — called "EnergyPLUS Home 4.0"— was built and entered in the competition. The energy balance and indoor comfort test reached the design value. This book with its eight chapters introduces the detailed project design and construction process and aims to serve as a reference for the construction of active buildings in China.

The first chapter introduces the overview and development of energy-saving buildings. It is the theoretical basis of the main part of the book, which includes the following chapters. The second chapter introduces the space concept of the project: how to use solar energy passively through design, using "box"

to develop an ensemble system. Closely combined with the space design is the energy system of the building, which is explained in the third chapter. The technical strategy includes active and passive measures, obtained after analyzing the climatic conditions of the location. The 11 sections of the fourth chapter introduce the main materials used in the building. The fifth chapter introduces the technical measures and equipment. Chapters 6 and 7 describe the prefabrication process and the quick construction process. The eighth chapter shows the performance and the data in the competition.

The materials and technical equipment used in the project are generally from domestic manufacturers. Their products and technologies are produced for energy-saving requirements, and these companies have leading positions in their fields. The book includes their information and URL links, to serve the readers who want to do more research.

目　录
C o n t e n t s

前言

1 主动式・被动式 / 1

2 设计・理念 / 7
   2.1 国际太阳能十项全能竞赛 / 8
   2.2 两届冠军作品的精华 / 9
   2.3 正能量房 4.0 的空间理念 / 11
   2.4 整体装配式："盒子"单位 / 16
   2.5 外围护结构设计策略 / 22

3 能量系统 / 29
   3.1 气候条件 / 30
   3.2 技术策略：主动式 + 被动式 / 34
    • 被动式节能 / 34
    • 太阳墙 / 37
    • 制冷和供暖系统 / 39
    • 新风系统 / 40
    • 主动式产能 / 41

4 材料 / 43
   4.1 外围护结构——透明构件 / 44
   4.2 外围护结构——不透明构件 / 48
   4.3 缓冲区——透明构件 / 52
   4.4 防水系统 / 56
   4.5 地板——福尔波亚麻地板 / 59
   4.6 卫浴设施 / 63
   4.7 一体化厨房 / 65
   4.8 推拉木门 / 66
   4.9 照明 / 68
   4.10 工程木 / 70

Preface

1 Active・Passive / 1

2 Design・Concept / 7
   2.1 International Solar Decathlon Competition / 8
   2.2 The essence of the two championship entries / 9
   2.3 Space concept of EnergyPLUS Home 4.0 / 11
   2.4 Assembled system: Box unit / 16
   2.5 Building envelope design strategy / 22

3 Energy System / 29
   3.1 Climate conditions / 30
   3.2 Technical strategy: active + passive / 34
    • Passive energy-saving / 34
    • Solar wall / 37
    • Cooling and Heating System / 39
    • Fresh air system / 40
    • Active energy production / 41

4 Materials / 43
   4.1 Building envelope: Transparent components / 44
   4.2 Building envelope: Opaque components / 48
   4.3 Buffer zone: Transparent components / 52
   4.4 Waterproof System / 56
   4.5 Floor material: Forbo Linoneum / 59
   4.6 Sanitary facilities / 63
   4.7 Integral kitchen / 65
   4.8 Sliding timber door / 66
   4.9 Lighting system / 68
   4.10 Engineered wood / 70

4.11 室外场地 / 71

5 技术 / 73
   5.1 制冷 / 74
   5.2 供暖 / 76
   5.3 热泵和缓冲水箱 / 77
   5.4 太阳能光伏板 / 79

6 预制 / 83
   6.1 钢主体承重结构 / 84
   6.2 轻钢结构 / 86
   6.3 门窗、侧墙 / 87
   6.4 顶部暗藏管线、内隔墙 / 88
   6.5 底部地暖与面层基层 / 89

7 施工 / 91
   7.1 组装 / 92
   7.2 缓冲区钢（木）结构安装 / 94
   7.3 屋面 / 95
   7.4 太阳墙 / 96
   7.5 缓冲区推拉扇 / 96
   7.6 室内终饰 / 97
   7.7 地板面层 / 97
   7.8 一体化厨房 / 97
   7.9 室外平台、凉亭和水池 / 98
   7.10 活动家具 / 98
   7.11 杂项 / 99

8 实测 / 101
   8.1 室内温湿度 / 102
   8.2 室内 PM2.5 值 / 106
   8.3 产能量和总能耗 / 106

附录一：同济大学——达姆施塔特工业
      大学联合赛队成员名单 / 108
附录二：正能量房 4.0 项目赞助名单 / 109
附录三：2018 中国国际太阳能十项全能
      竞赛比赛规则 / 110
后记 / 111

4.11 The site / 71

5 Techniques / 73
   5.1 Cooling / 74
   5.2 Heating / 76
   5.3 Heat pump and buffer tank / 77
   5.4 PV-Panels / 80

6 Prefabrication / 83
   6.1 Steel load-bearing structure / 84
   6.2 Light steel structure / 86
   6.3 Windows and sidewalls / 87
   6.4 Hidden pipeline, interior partition
       walls / 88
   6.5 Floor heating and floor base layer
       / 89

7 Construction / 91
   7.1 Reassembly / 92
   7.2 Steel and wood structure assembly
       in the buffer zones / 94
   7.3 Roof / 95
   7.4 Solar walls / 96
   7.5 Sliding elements in the buffer zones
       / 96
   7.6 Interior finishing / 97
   7.7 Flooring / 97
   7.8 Integral kitchen / 97
   7.9 Outdoor platform, pavilion, pool / 98
   7.10 Movable furniture / 98
   7.11 Miscellaneous / 99

8 Performance / 101
   8.1 Indoor Temperature and humidity
       / 102
   8.2 Indoor PM2.5 / 106
   8.3 Energy production and total energy
       consumption / 106

Appendix 1. Member list of Team Tongji
            University—Technical
            University of Darmstadt / 108
Appendix 2. Sponsor list of EnergyPLUS 4.0
            Home / 109
Appendix 3. The rules of Solar Decathlon
            Competition China 2018 / 110
Acknowledgement / 111

主动式·被动式
Active·Passive

1

从被动式到主动式——绿色节能建筑的未来

From passive to active – the future of green energy-efficient buildings

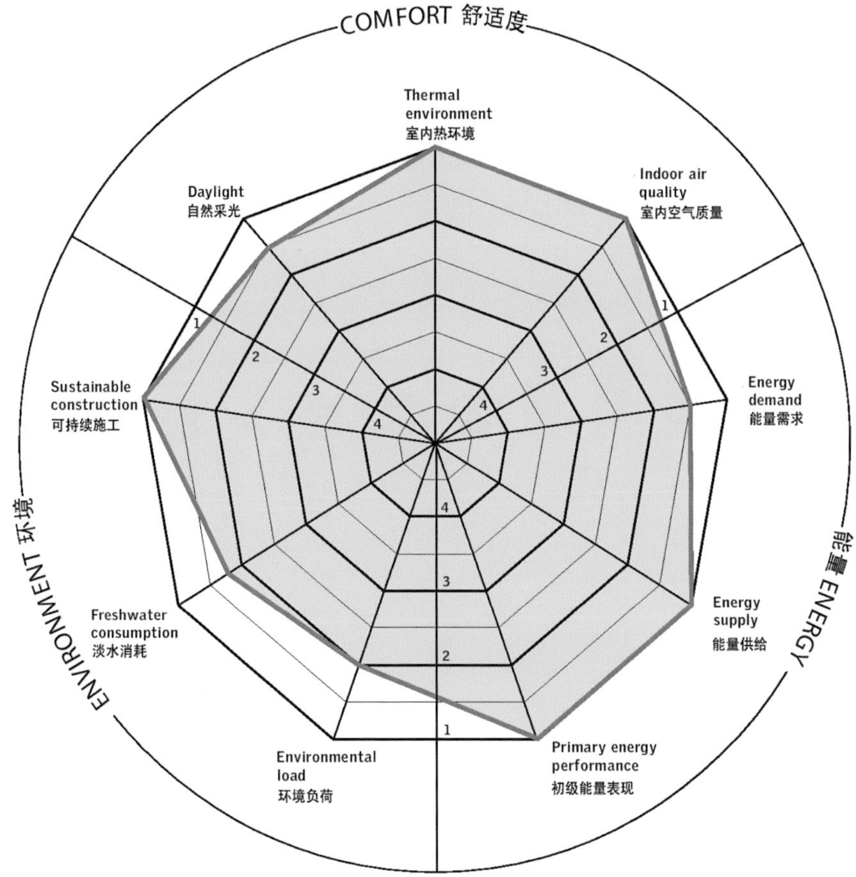

COMFORT 舒适度
Thermal environment
室内热环境
Indoor air quality
室内空气质量
Energy demand
能量需求
Energy supply
能量供给
能量 ENERGY
Primary energy performance
初级能量表现
Environmental load
环境负荷
Freshwater consumption
淡水消耗
ENVIRONMENT 环境
Sustainable construction
可持续施工
Daylight
自然采光
1
2
3
4

从世界范围内来看，建筑从 20 世纪 70 年代能源危机开始与能耗挂起钩来，至今为止可以说经历了四个发展阶段。第一个阶段可以称作觉醒阶段，20 世纪 70 年代的能源危机使得能源开始成为全世界需要共同考虑的问题。人们逐渐意识到，以牺牲生态环境为代价的高速文明发展史难以为继，耗用自然资源最多的建筑产业必须走可持续发展之路。

这个阶段延续到第一部建筑保温法的颁布（1977 WSVO），对建筑能耗控制开始有了具体的法律规定，节能建筑的评价和约束体系逐渐成形。20 世纪 80 年代，节能建筑体系逐渐完善，并在英、法、德、加拿大等发达国家广为应用，在此法规颁布前的建筑就有一些节能尝试，如在寒冷地区采取外墙保温措施，这个法规执行后建造的建筑明确地针对建筑外围护结构的保温隔热性能采取了一些改善性措施。

建筑保温法在 1984 年和 1996 年进行过两次修订，直至 21 世纪初期，被 EnEV

From a global perspective, energy-efficient buildings have experienced four stages of development. The first phase can be called the awakening phase. It started with the energy crisis of the 1970s which made energy use a major issue. People gradually realized that the conflict between economic growth and the environment is unsustainable. Thus the construction industry, which consumes the most natural resources, must follow the path of sustainable development.

This concern led to the Thermal Insulation Ordinance 1977 (1977 WSVO), which was the beginning of specific legal control for building energy consumption. As a result, the evaluation and restraint system for energy-efficient buildings gradually took shape. There were some improvements in the 1980s, with energy-saving attempts such as the external wall insulation measures in cold areas. Some buildings constructed after this regulation

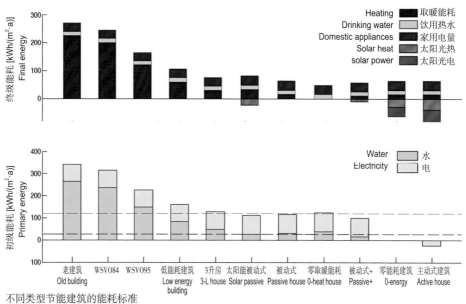

不同类型节能建筑的能耗标准
Energy consumption standards for different types of energy-efficient buildings

取而代之，这个阶段可以称作低能耗建筑阶段，主要体现在节能法规的不断完善与深化，在世界各国陆续颁布建筑节能法规，对新建与改建建筑的能耗要求逐年提高，建筑节能措施也不断拓展，一些有利于建筑节能的手段也在空间设计上得以体现，如中庭、双层立面等的运用。

被动式建筑理念出现在 20 世纪 90 年代初，到 21 世纪初期逐渐推广，1999 年年底德国的被动式建筑有大约 300 个，2000 年底就已经有大约 1000 个，而 2006 大约在 6000 ~ 7000 之间。虽然它不是能源标准，而是一个全面的高舒适度概念："被动式建筑是热舒适度（ISO7730）仅通过对新风的再加热或再冷却就可以满足，不需要使用循环空气，而新风的空气质量完全可以达到要求（DIN1946）"。这个时期对建筑能耗的要求也越来越高，一些概念层出不穷，除了被动式还有类似 3 升房、1.5 升房、零能耗建筑等，总的来说就是采取各种措施让建筑的能耗降低到极限。

被动式建筑尽可能通过被动的措施如热保温、通过温差进行热回收、被动地利用太阳能和室内热源等来实现热舒适度。被动式建筑要求有极好的保温效果，所有裸露于室外的建筑部位包括屋面、外墙、地下室顶板或建筑底板的保温系数需达到 0.15 W/(m² · K)，外窗必须有极好的保温性能，如三层中空玻璃窗（U=0.7 W/(m² · K)；$g \cong 50\% ~ 60\%$），带有高效热回收的可控的通风系统保证室内的空气质量，减少通风带来的热损失。

伴随着对低能耗的追求，还出现了另外一个呼声，即与其竭尽全力节能，不如想方设法产能，用充裕的可再生能量来平衡与弥补建筑的能耗。如果建筑产出的能量比自身消耗的还多，那么它的能量平衡就是正的，也就是正能量建筑。

要达到能量剩余，只能通过主动式，仅通过被动式节能是无法做到的。要利用建筑的屋面、立面产能，可以

have also included improvement measures for the thermal insulation of the building's external structure.

The WSVO was revised twice in 1984 and 1996 and was replaced by EnEV in the early 21st century. This can be called the low-energy building stage, which is mainly reflected in the continuous improvement and deepening of energy-saving regulations in countries around the world. Building energy conservation regulations have been issued, energy requirements and measures for new and renovated buildings have increased year by year. Means of building energy conservation have been reflected in space design, such as the atrium, double facade, etc.

The concept of the passive building appeared in the early 1990s. It was gradually popularized by the beginning of the 21st century. At the end of 1999, there were about 300 passive buildings in Germany. At the end of 2000, there were about 1,000, and in 2006, there were about 6,000~7,000. Although passive building is not an energy standard, it is a comprehensive concept of high comfort: "The thermal comfort of the passive building (ISO7730) can only be met by heating or re-cooling fresh air, without the need for circulating air, and the quality of the fresh air can be fully met (DIN 1946)". During this period, the requirements for building energy consumption became tougher, and new concepts emerged such as the 3-liter house, 1.5-liter house, zero-energy building, etc.

Passive buildings achieve thermal comfort through measures such as thermal insulation, heat recovery, use of solar energy and indoor heat sources. Passive buildings require excellent insulation. All components that are exposed to the outside, including roof, exterior walls, basement or floors, must have U-Value of 0.15 W/(m² · K). The exterior windows must have excellent insulation properties, e.g. the windows with triple glazing (U=0.7 W/(m² · K); $g \cong 50\% ~ 60\%$). A controlled ventilation system with efficient heat recovery ensures indoor air quality and reduces heat loss from ventilation.

利用太阳能、废水中的热能回收、生物能、风能等。此外，还要利用基地上和基地周边一切有利的条件，而最有效的还要利用建筑组群，形成产能和耗能的区域联网。一个比较有代表性的实例是2015年刚建成的法兰克福施派歇尔大街多层城市住宅（具体参见《建筑节能技术与建筑设计》，曲翠松著）。

主动式建筑是基于被动式建筑发展而来的，它的设计首先要遵循被动式建筑的设计原则，即降低建筑本身的能耗，再加上合理的产能措施。由于有了自身生产的可再生能量，主动式建筑在空间和造型设计上都更为灵活。这个发展和建筑领域技术和材料的进步密不可分，如太阳能光电产品和过去相比在性能和价格上都有极大的突破，再如地热

Beyond low energy consumption, there is another objective: instead of merely saving energy, a passive building can also include ways to produce energy. If a building produces more energy than it consumes, its energy balance is positive, that is an energy surplus building.

Energy surplus cannot be achieved only through passive energy-saving. To use the roof and facade of a building to produce energy, to use solar energy, heat recovery in wastewater, biomass, wind energy, etc., the building must take advantage of all the available conditions on the site and surroundings. The most effective way to do this is to group buildings to form a regional power network for energy production and consumption. A good example is the multi-storey urban residence of Speicher Street in Frankfurt,

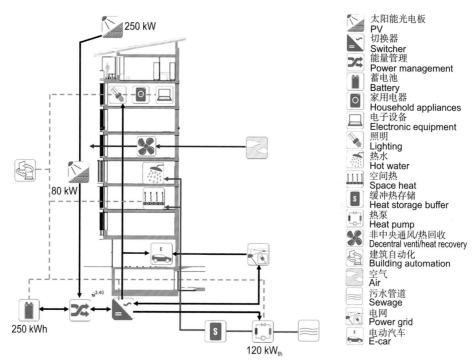

太阳能光电板
PV
切换器
Switcher
能量管理
Power management
蓄电池
Battery
家用电器
Household appliances
电子设备
Electronic equipment
照明
Lighting
热水
Hot water
空间热
Space heat
缓冲热存储
Heat storage buffer
热泵
Heat pump
非中央通风/热回收
Decentral venti/heat recovery
建筑自动化
Building automation
空气
Air
污水管道
Sewage
电网
Power grid
电动汽车
E-car

法兰克福施派歇尔大街多层城市住宅能量管理示意
Energy management of multi-story residential buildings on Speicher Street in Frankfurt

利用的开发也使得地热用于建筑更有吸引力。这些措施虽然会提高建筑初期的投资，但是从建筑全生命周期来看，主动式建筑是优于传统建筑的。随着建筑技术的发展，主动式建筑在未来将会越来越多地被运到各种类型的建筑中。

　　正能量房 4.0 是目前在我国建成的唯一一座主动式建筑，是同济大学和德国达姆施塔特大学在 2018 年中国国际太阳能竞赛中的共同参赛项目。从设计概念的形成，到方案深化、落实实施细节、寻找材料和产品，直至预制、比赛现场施工和比赛测评，项目历时两年半。

　　建筑立足于中国市场，采取主动式与被动式相结合的设计方法，通过主动式建筑的设计理念打造适合中国国情的高能效绿色建筑。在比赛期间建筑的产能大大高于自身能耗，并且舒适度极高。作为参赛作品，正能量房 4.0 的实验性和展示性较强，为主动式建筑在我国的后续发展提供了很好的参考。

completed in 2015 (see *Building Energy Saving Technology and Architectural Design*, Qu Cuisong).

The active building concept has been developed based on passive building. It follows the design principle of passive building to reduce energy consumption and takes suitable measures to produce energy. Active buildings are more flexible in space and design due to their renewable energy, thanks to the advancement of technology and materials in the field of construction. Solar photovoltaic products have made great breakthroughs in performance and price, and increased geothermal utilization has made geothermal energy more attractive for buildings, etc. Although these measures increase initial investment, they pay off over time in terms of lower energy costs. With the development of building technology, active buildings will have significant potential.

EnergyPLUS Home 4.0 is the only active building currently built in China. It is a joint project of Tongji University and Technical University of Darmstadt in Germany in the 2018 China International Solar Competition. The project lasted two and one-half years, including connectional design, development design, implementation of details, the search for materials and products as well as prefabrication, on-site construction, and competition evaluation.

Focused on the Chinese market, the project combines active and passive design methods to create an energy-efficient green building suitable for China's national conditions. During the competition, the building's produced much more energy than it consumed, and comfort is extremely high. As an experimental and demonstration project, EnergyPLUS Home 4.0 provides a good reference for follow-up development of active buildings in China.

设计·理念
Design·Concept

2

EnergyPLUS Home 4.0, 正能量房 4.0，是同济 – 达姆队参赛项目的名称。设计和项目实施都以实现"正能量"为目标。

EnergyPLUS Home 4.0 was the project title of the Team Tongji&TUDA and revealed the design and implementation goal.

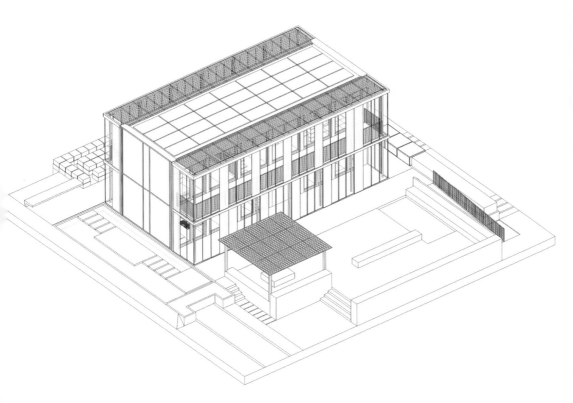

2.1 国际太阳能十项全能竞赛

国际太阳能十项全能竞赛是以全球高校为参赛单位的太阳能建筑科技竞赛，2002 年在美国华盛顿 D.C. 成功举办了首届太阳能十项全能竞赛，一直到 2018 年夏季在中国山东省德州市刚刚完成的我们所参加的这一届，已经走过了 16 年，分别在北美、欧洲、亚洲和南美洲举办了 10 届。Richard King 先生怎么都不会想到这项当年由他发起的赛事会在世界范围内产生如此重大的影响。这和所到之处各个国家政府、高校和企业共同的积极参与支持是分不开的，它也同时见证了人们对致力于保护我们的地球家园、建设绿色生态的生存环境的共识。

国际太阳能十项全能竞赛借助世界顶尖研发、设计团队的技术与创意，将太阳能、节能与建筑设计以一体化的新方式紧密结合，旨在设计、建造并运行一座功能完善、舒适、宜居、具有可持续性的太阳能住宅。太阳能十项全能竞赛虽然开列了十项比赛规则，但核心理念是太阳能，这是毋庸置疑的。纵观 2002 年第一届国际竞赛以来的发展，这项国际赛事引领了世界建筑节能设计理念，推动了建筑节能技术的发展，好的参赛作品在这两个方面必须具有前瞻性。

2.1 International Solar Decathlon Competition

The International Solar Decathlon Competition is a solar building technology competition with global universities as the participating units. In 2002, the first SDC was held in Washington DC, USA. Including the competition in the summer of 2018 in Dezhou, China, 10 sessions have been held in four continents. Richard King would never have imagined that the event he initiated has had such an impact. This success required the active participation of governments, universities, and enterprises. It also illustrates the public's interest in protecting our earth and to building a green ecological environment.

The SDC combines the technology and creativity of the world's top R&D and design teams to integrate solar energy, energy conservation, and architectural design. The goal is to create a functional, comfortable, and sustainable solar home. Although there are ten rules in the competition, the core issue is solar energy. Since it began in 2002, the SDC has been a leader in energy-saving design and HAS promoted the innovation of energy-saving technology. Good entries must be forward-looking in both areas.

太阳能十项全能竞赛 2007 年冠军作品，达姆施塔特工业大学
The winner entry of SDC 2007, TU Darmstadt

2.2 两届冠军作品的精华

德国达姆施塔特工业大学一共参加了两次在美国举办的国际太阳能竞赛，分别是 2007 和 2009 年，在已故的黑格尔教授的带领下，团队连续获得了这两届比赛的冠军。当年的比赛规则与现在的不太一样，比如 2007 年的规则里建筑设计、工程技术和能量平衡项的分数权重都比其他项目要高。2007 年的冠军作品在这些单项里都获得了最高分。设计关键点：

- 规则与紧凑形体
- 简洁明了的平面布局与功能分区
- 南侧缓冲区
- 太阳能建筑一体化

尤其是把太阳能光伏构件整合到南侧立面的木制遮阳百页上的做法，在当年还属于首创级，为项目的建筑设计理念部分赢得了创新点。此外，在建筑的内装饰材料中加入了缓冲与相变材料，在白天尽量吸收能量，存储起来供建筑在夜晚室外温度降低时使用。

2.2 The essence of the two championship entries

The Technical University of Darmstadt in Germany participated in competitions in 2007 and 2009 in the United States, led by Professor Hegel (†) , and won both. The rules were different than the current competition, e.g. the scores of architectural design, engineering technology and energy balance in 2007 were higher than those of other categories. The 2007 champion team won the highest score in each of these categories. Key points of their design concept:

- Regular and compact shapes
- Concise floor layout and functional partitioning
- Southern buffer zone
- Solar building integration

The practice of integrating solar photovoltaic components into the wooden sunshade on the southern facade was pioneering, and with a PCU cooling concept it won innovation points for the design of the project.

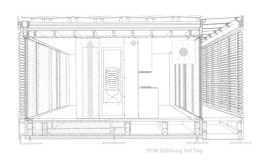

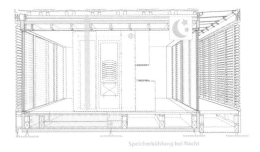

太阳能十项全能竞赛 2007 年冠军作品，纵剖面，内部相变材料制冷
The winner entry of SDC 2007, Section: PCM cooling concept

相比之下 2009 年的比赛则更关注能量的产出。建筑设计和工程技术的得分都被下调至 100 分，而入网电量一项却上调至满分 150 分。这一次达姆施塔特大学队的策略是最大限度地利用建筑的外表面产能，"德国队专注于产能过剩，把建筑所有的维度能用的都用足，每一块可利用的表面都覆盖了太阳能光伏板"。建筑看上去像是穿了一身闪亮的黑色盔甲，光是入网电量这一项就为赛队赢得了满分，总分数遥遥领先于其他赛队。

设计关键点：
- 规则与紧凑形体
- 平面布局与功能分区更丰富，形成跃层空间
- 立面太阳能产能最大化，南侧封闭，北侧开敞
- 工业制造，立面构件模数化

The 2009 competition focused more on energy output. The scores for architectural design and engineering techniques were all lowered to 100 points, while the net power was increased to 150 points. This time the strategy of the TU Darmstadt team was to maximize the energy production of the building surface. "The German team focused on energy surplus and was able to use all the dimensions of the building, with every available surface covered by PV panels." The building appears to be wearing shiny black armor, and the net power won the full score. The total score was far ahead of other teams. Key points of its design concept:
- Regular and compact shapes
- Layout and functional partitioning are more complex, providing a mezzanine
- Maximizing solar energy production on the facade, closed on the south side and open on the north side
- Industrial manufacturing, modularization of facade components

太阳能十项全能竞赛 2009 年冠军作品
The winner entry of SDC 2009, TU Darmstadt

2.3 正能量房 4.0 的空间理念

　　EnergyPLUS Home 4.0，中文译名"正能量房 4.0"，是同济－达姆队参赛项目的名称。它在赛队提交参赛申请计划时就已经形成了，其后所进行的设计以及项目的实施都是为实现"正能量"所进行的。目标很明确，具体的做法则要利用现有的条件，认识中国市场与产品的局限，随时进行必要的调整。

　　2016 年的春季是令人愉悦的，设计任务对中国和德国学生来说同样富有挑战性。我们既要学习以往成功的经验，又要在此基础上有所提高，要把技术和认知都已经进步了的时代性鲜明地体现出来。由于竞赛项目以学生为主，在各自教师和教研室团队的指导下，45 名德国学生和 6 名中国学生以 2018 年中国国际太阳能竞赛的任务为毕业设计题目以及 6 名中国研究生同学作为学期设计课题一起参加了方案设计，经过 4 个月的设计，中德双方组成评委，评选出优秀作品并将最能体现技术发展趋势和最有市场适应性潜力的优胜方案作为深化设计和参赛项目的发展基础。

2.3 Space concept of EnergyPLUS Home 4.0

EnergyPLUS Home 4.0 was the project title of the Team Tongji&TUDA and revealed the design and implementation goal. The specific approach takes advantage of existing conditions, in light of the limitations of the Chinese market and products.

The spring of 2016 was enjoyable, and the design task was challenging for both Chinese and German students. We must learn from the experiences of the past, improve and take the advantage of the developed technologies. Under the guidance of their respective teachers and research teams, 45 German students and 12 Chinese students participated in the concept design phase of the project. After four months, a jury formed from the Chinese and German sides has selected the outstanding works, which were the basis for the developing design of the entry project.

南立面　South elevation

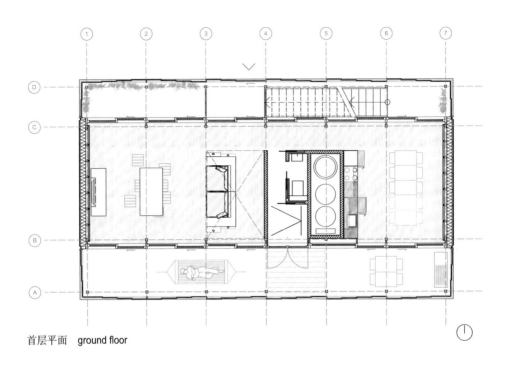

首层平面　ground floor

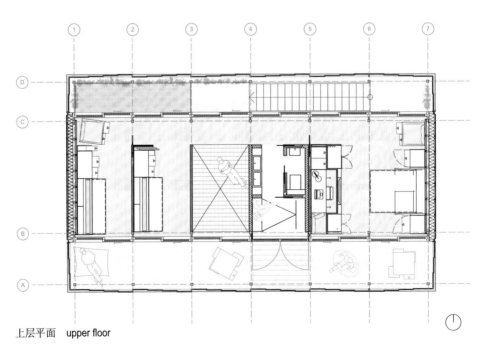

上层平面　upper floor

- 均质性

这个入选方案继承了两届获奖作品的优点，秉承了德国建筑在 20 世纪初期形成并延续至今的现代主义风格，在空间组织和形体构成上坚持简洁、规则、紧凑的特征并加以拓展，不再强制性定义单独空间，除了必要的服务和设备空间外，其他室内空间是均质的，均可根据使用需要和用户偏好自由分隔、组合和置换。

建筑的形体是一个二层高的长方体，每层都由中间的功能区（使用空间＋技术服务核心空间，此区域为气候控调区）、南北两侧的缓冲区组成。缓冲区为非气候控调的外廊或外阳台，在北侧外廊还安装了一部楼梯，因而也具有交通空间的功能。

- Homogeneity

This selected plan inherited the advantages of the previous two award-winning German projects. It keeps the modernist style and insists on simple, regular and compact features in its spatial organization and form composition. Furthermore, it abandons the compulsory definition of single spaces. Apart from the necessary services and equipment space, other indoor spaces are homogeneous and can be freely separated, combined and replaced according to the needs of use and user preferences.

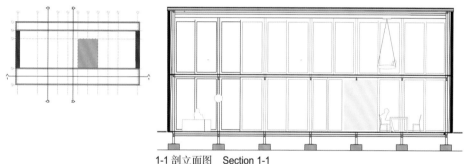

1-1 剖立面图　Section 1-1

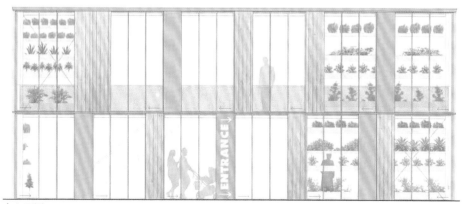

南立面图　North elevation

气候控调区内的空间是一个开放空间，建成后唯一不能更改的部分是服务性空间：设备间和卫生设施。底层技术服务核心部分包括卫生间，洗漱台，厨房和配电、空调、热水的技术用房；二层的相同位置是相似的服务空间（卫生间、浴室、洗衣间）。这个紧凑的服务单元也整合了上下贯通的各类技术管道井等。除此以外的其他空间都可以任意分隔，如底层可以分成起居空间和餐饮空间，这些又可以灵活定义成游戏空间、会议空间、健身休闲空间等。较大的起居空间也可以隔出一间客房，作为卧室使用。二层的大空间同样可以用作各种功能，在居住建筑中可用作带工作室或更衣室的主卧、次卧、工作室、儿童房或健身房等。

The building is a two-story cuboid. Each floor consists of a functional area in the middle (living space + technical serve core), the climate control zone, and a buffer zone on the north and south sides (outer gallery or terrace), the non-climate control area. In the northern outer gallery is a staircase, so that this area functions as part of the circulation.

The space in the climate control zone is an open space. The only units with fixed functions are the service cores: equipment room and sanitation. Various technical pipe wells are integrated in those two compact service units. All other spaces can be freely separated, combined or switch the original function to the new one. For example, the first floor can be divided into living space and dining space, which can be flexibly defined as communication space, meeting space, fitness or leisure space. Larger living space can also be separated into two rooms, and one of them can be used as a bedroom. The large space on the second floor can also be used for various functions. It can be separated into a master bedroom with a studio or a dressing room, a second bedroom, a studio, and a children's room or a gym in a residential building.

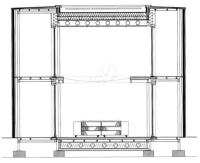

2-2 剖面
Section 2-2

房间内部材料选用示例
Material sample for private rooms

中庭做法示例
Option for the atrium

- 缓冲区

正能量房 4.0 的理念以被动式利用太阳光热为节能策略核心，而缓冲区则是一个有节能产能作用的功能区。它拓展了生活空间，将室内外融为一体，更为现代家庭提供了一种崭新的沟通平台。

缓冲区位于组装好的箱体的南北两侧，由一层轻质透明和可推拉移动的构件组成。南侧缓冲区在夏季是一个敞开的空间——处于阴影区的室外平台和阳台，而在冬季，通高的推拉扇关闭，白天直射的阳光将这里变成一个温暖的可以供人活动居留的空间，同时它也阻止室内空间直接与室外寒冷环境接触而散失热量。简言之，这是一个现代的、二层的阳光房，简洁、明亮，串联各个功能单元，为家庭成员提供了互动的空间。

位于北侧的缓冲区和南侧缓冲区一样，也为与其相邻的主体建筑提供了一道直接与室外接触的屏蔽，阻止冬季建筑内部的热量直接向室外传递。除此之外，由于建筑的主入口位于北侧，也出于更好地满足比赛展示参观需要的考虑，在这个参赛项目里选择将一部连接上下两层空间的直跑楼梯放置在北侧缓冲区，既展示了缓冲区，同时也可以作为交通空间服务于建筑主体的功能。

无论位于建筑主体的哪一侧，缓冲区的外层结构对处于缓冲区内的外围护结构都起到天气防护的作用，以减少主体外墙的维护，延长其使用寿命。

- Buffer zone

The concept of the project takes the passive use of solar heat as the main energy-saving strategy, while the buffer zone is a functional area with energy-saving capacity. It expands the living space, connects inside and outside, and provides a new communication space for the modern family life.

There are buffer zones on both sides of the assembled boxes, using lightweight, transparent, and sliding elements as the outer skin. The south buffer zone is an open space in the summer, providing outdoor platforms in shadow. In the winter, the floor-high sliding elements are closed, the sunlight during the day falls in, and it turns into a solar room for people to stay. This modern and bright solar space connects the functional units and provides space for interactive activities of the uers.

The northern buffer zone provides a contact shield between the main building and outdoor space, preventing the heat transfer in winter. The main entrance of the building is located on the north side, and to meet the needs of the exhibition, the project has placed a straight running staircase connecting the upper and lower floors in the northern buffer zone, which serves as part of the circulation.

The outer structure of the buffer zone acts as weather protection for the building facade. It can reduce the maintenance of the building envelope and extend its life span.

  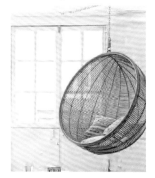

南侧缓冲区吊挂家具示意　Examples for hanging furniture in the buffer zones on the south side

2.4 整体装配式："盒子"单位

　　合理的装配系统可以提高建筑效率。如果大部分组件甚至整体区块都是在工厂里组装完成的，那么花费在工地现场的时间和工作量将大大减少。装配式的优点显而易见，这么做长远来看可以降低成本，但更重要的是工地上大部分湿作业被取代了，改善了建筑环境污染。选择装配式建造体系本身并没有创新之处，但是如何提高装配化程度，尽可能做到整体式装配以缩短工地建造时间和节约成本，这方面在整个世界范围内都还有很大的提升空间。
　　我国装配式建筑的发展和德国这样的工业化国家相比还处于较低的水平，选择装配式在项目实施中的难度是可以预见

2.4 Assembled system: Box unit

The assembly system of EnergyPLUS Home 4.0 is based on the "box", which will be prefabricated and installed with up to six surfaces in the factory. The box unit has a frame structure with all non-load bearing walls. The size of the frame is adjustable and determined by factors such as the processing, transportation, and function requirements. Depending on their positions, the boxes have walls, roofs, or floors. Each box is different; all its components must be well designed for the prefabrication.

Within the skeleton and enclosure of the box, the necessary "blood vessels" and

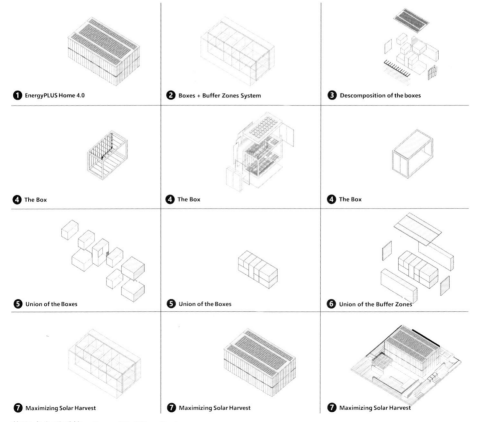

❶ EnergyPLUS Home 4.0　　　❷ Boxes + Buffer Zones System　　　❸ Descomposition of the boxes

❹ The Box　　　❹ The Box　　　❹ The Box

❺ Union of the Boxes　　　❺ Union of the Boxes　　　❻ Union of the Buffer Zones

❼ Maximizing Solar Harvest　　　❼ Maximizing Solar Harvest　　　❼ Maximizing Solar Harvest

装配式盒子系统　Assembled Box System

的，但这正是太阳能国际十项全能竞赛这样的项目之所以存在的关键点，它鼓励参与者的创新、挑战难题。

正能量房 4.0 的装配式是以"盒子"为单位的，换言之，空间的六个面都是在工厂预制和安装的。"盒子"单位是一个框架结构，所有的墙体均无需承重，这个框架的尺寸可以调整，根据加工、运输和使用功能等需求综合而定。这些盒子根据所在的位置有或者没有一些墙体、顶板或者底板。每个盒子都不相同；其所有组件都必须精心设计。

"盒子"的骨架和围护之内还要置入必须的"血管"和"呼吸道"组织，以达到内部空间的温度调控的目的，满足舒适度和节能需求。这些必须在设计时统一考虑，在预制时尽可能安装进去，以减少工地现场施工的工作量。

两层的参赛房屋展示了所有可能的围护类型，上下两层盒子之间的连接方式也得以体现。

"respiratory tract" must be installed to control the interior climate to meet the comfort and energy-saving requirements. These must be considered in the design, and installed as much as possible during the prefabrication.

The two-story house demonstrates all possible types of the enclosures, also the connections of the upper and lower boxes.

12 个各不相同的盒子单元
12 box units with 12 options

- 单体的使用灵活性

标准化的盒子数量可根据需要进行增减，房屋轴线数量的改变胡不会对总体概念产生影响。并且即便在确定了盒子的总数后，其组合起来的内部空间仍有极大的灵活性，这为家庭成员随着时间变化的情况提供满意的解决方案。

- Combination Flexibility

The number of standardized boxes can be increased or decreased as needed, and the change in the number of axes does not affect the overall concept. Even after determining the number of boxes, the combined internal space is still extremely flexible, which gives an option when the family members change over time.

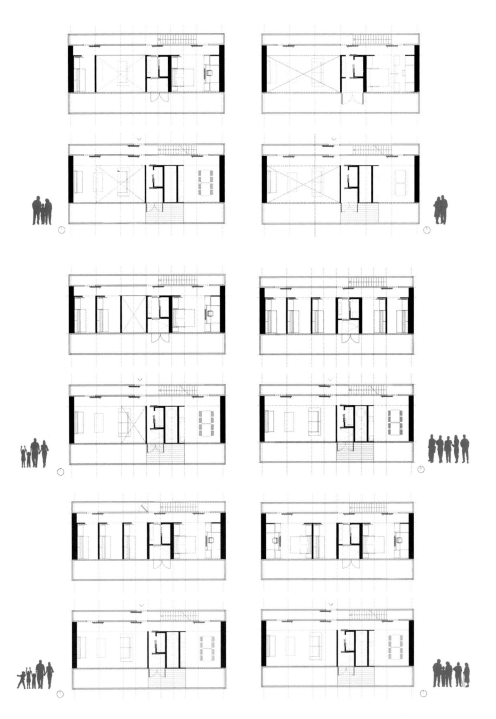

- "盒子"的结构

盒子采用框架结构，无论采用什么材料，它的外墙和内隔墙都是非承重的。这大大提高了空间的灵活度。盒子的承重结构可以采用钢材、木材和钢筋混凝土等。考虑到我国建筑法规、建筑工业的发展状况以及装配式体系要能够市场化推行，参赛的最终实施方案采用了钢结构。柱子：63mm × 160mm 槽钢，底部梁：73mm × 200mm 槽钢，顶部梁：63mm × 160mm 槽钢。

盒子的墙体还可以根据需要承担除分隔以外的功能，如悬挂其他构件，变成小空间的扩展功能墙。原设计中卧室的床即为挂在墙上的可折叠床，这个想法在实施时由于找不到合适的家具供应商而取消了。但入口中庭处的功能墙留了下来。

这面墙体按照楼梯踏步宽度模数在内部设了12根钢柱与结构框架梁焊接，立柱上按照楼梯踏步高度模数钻螺栓孔，这样即可根据需要在指定部位连接

- Structure of the box

Regardless of the material chosen, the box is framed so that its walls are non-load bearing, whether it is the outer wall or the inner partition wall. This greatly increases the flexibility of the space. The load-bearing structure of the box can be made of steel, wood or reinforced concrete. Considering China's building regulations, the development of the construction industry and the assembly system to be market-oriented, the final implementation of the project adopted a steel structure. Column: 63mm×160mm channel steel, bottom beam: 73mm×200mm channel steel, top beam: 63mm×160mm channel steel.

The partition walls can have other functions, such as hanging other components, becoming a functional wall of a small space. In the original design, the bed in the bedroom was a foldable bed hanging on the wall. The idea has been canceled due to the failure of finding a suitable furniture supplier. But the functional wall at the entrance with the atrium has remained.

This wall is welded with 12 steel columns to the structural frame beams according to the stair tread

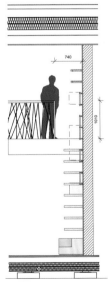

功能墙剖面
Section of the functional wall

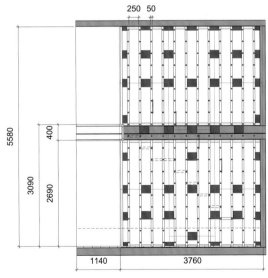

功能墙钢结构立面
Steel structure of the fuctional wall

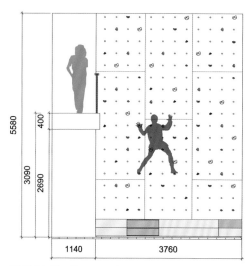

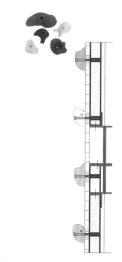

攀岩墙立面　Elevation of the climbing wall

攀岩墙细部　Detail Climbing Wall

螺杆，用以支撑踏步板、搁物板及攀岩点等。在比赛时由于是夏季且出于活动展示考虑，多功能墙以攀岩墙的形式体现。

width modulus. The column is drilled with bolt holes according to the tread height so that the screw can be connected at a specified location as needed to support treads, shelves, or climbing points. Due to the summer season of the competition and for the event display, the multi-functional wall is embodied in the form of a climbing wall.

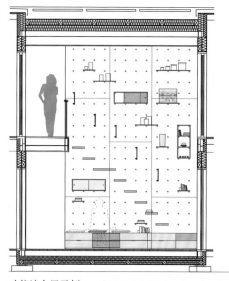

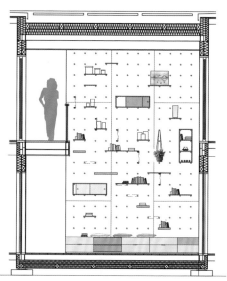

功能墙布置示例一　Option 1 of the functional wall

功能墙布置示例二　Option 2 of the functional wall

- "盒子"的组合方式

带有完整系统的"盒子"可以根据用户和场地的需要进行组合，根据承重体系的不同配置，使用"盒子"进行拓展，可以建成独栋、低层、多层和高层建筑。

- Combination of the boxes

The box with its complete system can be combined according to the needs of the user and the site conditions. Depending on the configuration of the load-bearing system, the boxes can be expanded into single-family, low-rise, multi-story and high-rise buildings.

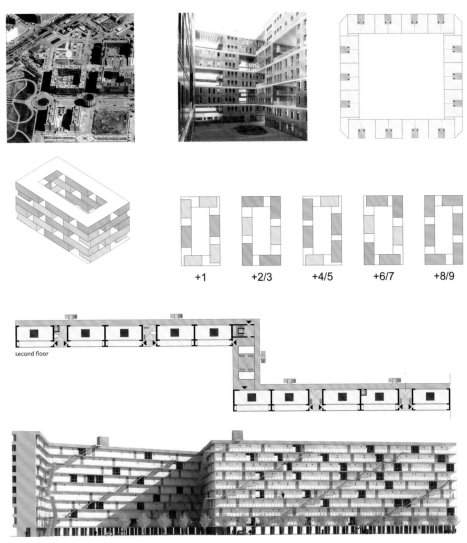

"盒子"集合建筑示例   Example project of corwbined boxes

2.5 外围护结构设计策略

盒子单元需要一个明确的外围护结构设计策略，既要满足通透性，保证良好的采光，又要保证在寒冷的冬季及酷热的夏季建筑内部与室外之间通过外围护结构产生热传递而损失的能量最低。

最初的方案设计是将盒子直接与外界接触的南北两个面全部采用落地门窗，使建筑完全通透。这样做的前提是窗的保温隔热性能足够好，好到可以和实墙接近。在德国，目前高档的节能建筑可以选用三腔四玻的门窗产品，其 U 值可以达到 $0.46W/(m^2 \cdot K)$，总厚度 30cm 的外保温砖墙的 U 值大约为 $0.32W/(m^2 \cdot K)$（17.5cm厚砖墙加聚氨酯混合保温系统），二者的保温性能已经相当接近。

虽然顶端的保温窗其保温性能极佳，但与轻质的不透明保温材料相比还有一定距离，比如一种基于硅酸盐的矿物质保温材料 CALOSTAT，热保温系数仅为 0.019，即使用 5cm 厚的该保温材料，墙体的 U 值即可达到 $0.38W/(m^2 \cdot K)$。三腔四玻的保温窗价格不菲，且在国内还没有相应产品。况且在北侧安装的窗主要用于满足采光需求，并无南侧窗在冬季被动利用太阳能以降低采暖需求的功效。因此项目组在进行项目的深化设计时，对于南北立面的处理做了进一步的研究和论证并进行了修正。具体的措施是在保留南侧立面全部采用落地门窗的同时优化盒子的北侧立面，将窗墙比提高 50%。由于南侧每个盒子开间一分为二安装两扇大窗，其中一扇固定，另一扇推拉；北侧与之相对应，一半为实体墙，另一半为平开窗。

2.5 Building envelope design strategy

The box unit requires a good envelope design strategy that ensures the sufficient daylighting and the lowest energy loss in winter and summer.

The original concept had completely transparent envelopes on both north and south sides, using the sliding windows and doors in floor-height. This only works when the thermal insulation of the window is as good as those of the solid wall. In Germany, high-end energy-saving buildings can use 4-glazing window products, which can reach the U-Value in $0.46W/(m^2 \cdot K)$, quite close to the U-Value of the thermal insulation brick wall with a total thickness of 30cm (about $0.32W/(m^2 \cdot K)$, 17.5cm brick wall, polyurethane mixed insulation system).

Although the thermal insulation of the top insulation windows is excellent, it is still not as good as the lightweight opaque insulation material. CALOSTAT, a silicate-based mineral insulation product has a heat transfer coefficient of only 0.019. For example, 5cm of CALOSTAT can achieve a U-Value of 0.38 $W/(m^2 \cdot K)$. The 4-glazing insulation window is very expensive, and there is no such product in China. Moreover, the windows on the north side have no function of utilizing solar energy in the winter like the windows on the south side. Therefore, the project team revised the facade design during the project development design phase and reduced the window-to-wall ratio by 50% on the north side while keeping the south side fully transparent.

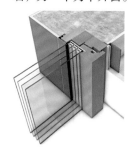

德国 Pazen 门窗生产的三腔四玻玻璃窗
3/18/2/18/2/18/3，总体 U 值 0.46W/($m^2 \cdot K$)
4-glazing window (3/18/2/18/2/18/3) produced by Pazen doors and windows in Germany. U-Value 0.46 W/($m^2 \cdot K$). Source: https://www.glaswelt.de/Archiv/Newsletter-Archiv/article-551756-112170/pazen-fenster-mit-uw-046-und-4-fach-iso-bekommt-pha-zertifikat-.html

• 节能效果论证

与原设计方案相比，北侧窗墙比提高 50% 的节能效果如何？项目组邀请了瑞士的绿色建筑专家加入顾问组，并通过节能模拟计算进行了前后方案的能耗对比。

具体的计算方式是按照瑞士现行的 Minergie 标准对新建建筑能耗要求计算的［Minergie：瑞士低能耗建筑标准，具体要求参见《主动式建筑：从被动式建筑到正能效房》（ *Vom Passivhaus zum Energieplushaus*)，［德］曼弗雷德·黑格尔 (Manfred Hegger) 编著，曲翠松译］。计入能耗面积 165m²，建筑与外界接触的六个表面计算情况如下表，建筑外围护结构面积系数 A/AE 取 2.20。

• Energy-saving effect demonstration

Compared to the original design, how is the energy-saving effect improved by the window opening reduction? The project team invited a Swiss green building expert to join the advisory group and calculated the energy consumptions of both options.

The calculation was according to the energy consumption requirements for the new buildings of Minergie standard (Minergie is a registered quality label for new and refurbished low-energy-consumption buildings in Switzerland). The energy consumption area is 165m². The calculation of the building surface is as follows. Building envelope area factor A/AE takes 2.20:

面积Area (m²)	室外 Outside	不取暖Unheated		土壤Soil		取暖Heated	总和Sum	
		without b-factor	with b-factor	without b-factor	with b-factor		without b-factor	with b-factor
屋面Roof	82.3	0.0	0.0	0.0	0.0		82.3	82.3
墙 Wall	196.8	0.0	0.0	0.0	0.0	191.9	388.7	196.8
地 Ground	82.3	0.0	0.0	0.0	0.0	0.0	82.3	82.3
总和 Sum	361.3	0.0	0.0	0.0	0.0	191.9	553.2	361.3

屋面、墙体及门窗面积，各建筑部位的保温性能计算值如下表：

Roof, wall, door/window area, the insulation performance of each building part as follows:

	屋面Roof	墙体 Wall				地板Floor	合计Sum
面积Area(m²)		N	E	S	W		
不透明部分Opaq.part	82.25	95.95	32.19	95.95	32.19	82.25	420.8
窗/门Window/Door		51.1		81.29			132.4
总和 Sum	82.25	147.05	32.19	177.24	32.19	82.25	553.2
Aw/A (%)		34.75		45.86			23.9
FS1		0.94		0.59			
FS2		0.76		0.58			
FS3		1.00		0.99			
FS		0.71		0.34			

门/窗面积与计入能耗面积之比 Ratio of window/door area to energy consumption area: 80.5 ‰

Aw/A: 门/窗面积与围护面积之比 Ratio of window/door area to enclosure area

FS1-3 面积权重阴影系数Area-weighted shading factors

FS1: 水平方向Horizon; FS2: 悬挑部分overhang; FS3: 侧面遮阳板side panel; FS = FS1 × FS2 × FS3

所有区域内门/窗面积与计入能耗面积之比Ratio of door/window area to energy consumption area in all zones: 80.5 ‰

在考虑了各个部位可能存在的线性和点式冷桥的基础上，计算出通过建筑物围护结构热传递产生的热损失：

Calculate the heat loss caused by the heat transfer through the building envelope, the linear and point cold bridges considered:

建筑部位 Component	编号 Code	朝向 Orientation	削减系数b Red. factor	U [W/(m²·K)]	面积 (m²) Area	热损失(MJ/m²) Heat loss
平屋顶 Roof	A1	天空Sky	1	0.12	82.3	20.06
墙体 Wall	B1	东 E	1	0.15	32.2	9.81
墙体 Wall	B1	西 W	1	0.15	32.2	9.81
窗 Window	G	北 N	1	0.83	51.1	51.12
窗 Window	G	南 S	1	0.83	81.3	63.68
墙体 Wall	G	北 N	1	0.15	44.9	8.11
墙体 Wall	G	南 S	1	0.15	14.7	2.12
底板 Ground	C3	室外Ouside	1	0.12	82.3	27.15
合计 Sum.						191.86

注：1. 墙体的 U 值按照目前瑞士外墙建造标准选取；2. 窗的 U 值按照三层中空窗选取。
Notes: 1. The U-Value of the wall is selected according to the current Swiss construction standards; 2. The U-Value of the window is selected according to the triple glazing window.

同时需要核算的还有通过门窗而产生的太阳能的热量，这个数值在不同月份里对建筑能耗产生不同影响。在计入建筑构件热存储能力和预设地板辐射取暖温度的基础上，得出建筑的全年能量平衡。

The solar heat generated through the doors and windows must be calculated. This value has different effects on building energy consumption in different months. The annual energy balance of the building is obtained by preset of the floor radiant heating temperature, with the thermal storage capacity of the building components included in the calculation:

类 型 Type	H (W/K)	QT (MJ/m²)	Qv (MJ/m²)	Qi (MJ/m²)	Qs (MJ/m²)	hg -	Qh (MJ/m²)	Qh,li-M (MJ/m²)
住宅/独立住宅 /2层 Residential/Single Family house/2 storeys	98.6	258.3	74.2	74.4	231.1	0.6	147.9	173.5
合计 (基本要求) Sum. (primary demands)		258.3					147.9	173.5

QT 热传导损失Transmission heat loss
Qv 通风热损失Ventilation heat loss
Qi, Qs 内部以及太阳能热获得
    Internal and solar heat gains
H 该类型的比传热系数
    Specific heat transfer coefficient of the type

hg 热获得利用率 Utilization rate of heat gains
Qh 取暖需求 Heating demand
Qh,li-M Minergie对Qh的最低值限制(0为建筑改建)
    Minergie limit for Qh (0 for renovation)
Qh,li 取暖需求限制Limit value for heating demand

同样的计算应用于三个模型，此处表格中采用的数据是最终实施方案模型。另外两个模型分别是原设计方案，即北侧全部采用落地窗，以及北侧完全不开窗。由于北侧完全不开窗将极大影响建筑的采光和风格，此模型仅作为比较用。计算结果证实了实施方案比原方案在能量平衡上的优越性。

The same calculations were applied to the three models. The data in the table here are from the final implementation model. The other two models are the original design and the model with no windows on the north side which is only used for comparison purposes. The calculation results confirmed the superiority of the implementation design in energy balance over the original design.

北侧外墙 50% 开窗率模型的月能量平衡表
Monthly energy balance table of the 50% window opening rate model of the north side outer wall

月份 Month	一月 Jan.	二月 Feb.	三月 Mar.	四月 Apr.	五月 May	六月 Jun.	七月 Jul.	八月 Aug.	九月 Sept.	十月 Oct.	十一月 Nov.	十二月 Dec.
持续时间 (h) Time constant	232	232	232	232	232	232	232	232	232	232	232	232
热损失 Q_t Heat loss	58.5	49.4	40.7	29.4	11.2	6.5	3.2	3.2	10.7	24.1	43.3	52.4
热获得 Q_g Heat gain	18.2	21.8	28.4	28.2	30.6	30.7	32.3	32	27.9	23.3	16.7	15.4
得热失热比 $g=Q_g/Q_t$	0.311	0.442	0.698	0.958	2.730	4.731	10.170	10.090	2.616	0.967	0.386	0.294
利用率 Utiliz. degree	1.000	1.000	0.999	0.961	0.366	0.211	0.098	0.099	0.382	0.957	1.000	1.000
利用的得热 Used profits	18.2	21.8	28.4	27.0	11.2	6.5	3.2	3.2	10.7	22.3	16.7	15.4
取暖需求 Heating demand	40.3	27.5	12.3	2.3	0.0	0.0	0.0	0.0	0.0	1.8	26.6	37.0

注:能量 EnergyMJ/m² 热损失: 热传递和通风热损失Heat loss: Transmission and ventilation losses

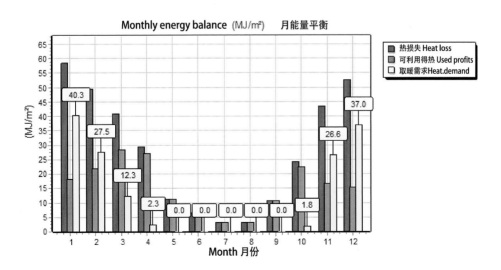

• 采光效果论证

通过大面积开窗达到现代建筑所表现的通透性，同时却不会因为增大了玻璃面积而影响建筑的能耗是正能量房4.0的本质所在。实施方案基于目前的建筑产品所能做到的能效极限做了修正，将北侧的开窗面积缩小50%，这在能量表现上有了提升并降低了建筑的总体造价，但是通透性也打了折扣。虽然这对于居住建筑来说在功能上不一定是缺点，但是是否能达到自然采光的要求，也需要经过论证。

首先通过三维建模导入模拟软件进行论证，具体设定如下（DIALux 4.12）：

• Daylight effect demonstration
Reducing the window opening rate on the north outer wall by 50% has improved the energy performance and reduced the overall cost of the building, but the transparency of the building also has been lowered. Although this is not necessarily a disadvantage for the building function, it still needs to be approved for daylight sufficiency.

First, set up the 3D modeling, then import it into the simulation software DIALux 4.12 for the demonstration. The specific parameter settings are showing below:

地点 Location：
中国山东省德州市 Dezhou, China
116E，37N，GMT + 8

时间抽样 Time samples：
春季中午 Spring noon，12:00/2017/3/21
春季傍晚 Spring evening，5:30/2017/3/21
夏季中午 Summer noon，12:00/2017/6/22
夏季傍晚 Summer evening，6:30/2017/6/22
冬季中午 Winter noon，12:00/2017/12/22
冬季傍晚 Winter evening，4:00 /2017/12/22

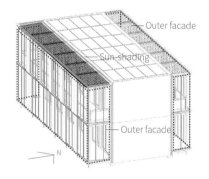

天气参数设置：
晴天和阴天：程序默认
计算表面高度：每层地板上方0.75米
材料：玻璃，单层，透明度：92%
（外围护窗户：三层玻璃，缓冲区推拉窗：双层玻璃）

遮阳：透明度：5%
缓冲区外层推拉窗状态：
春天，夏天：开放，冬天：关闭
可伸缩的遮阳状态：
晴天：阴影，阴天：阴影
位置：南侧缓冲区的玻璃顶下

Weather parameter setting:
Sunny and cloudy: program default setting
Calculate surface height: 0.75 meters above each floor
Materials: Glass, single pane, transparency: 92%
(External window: triplr glazing glass, sliding window in buffer zone: double glazing)
Shading: Transparency: 5%

Buffer zone outer sliding window state:
Spring, summer: open
Winter: closed
Retractable shading elements state:
Sunny day: shadow
Sunny day: shadow
Cloudy: shadow
Location: under the glass roof of the buffer zone in the south side

底层起居室北侧　North side, living room, 1F

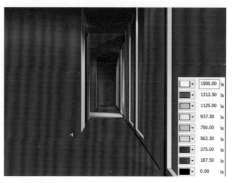
二层主卧北侧　North side, master bedroom, 2F

二层主卧室南侧　South side, master bedroom, 2F

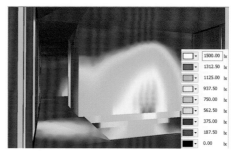
二层走廊　Corridor, 2F

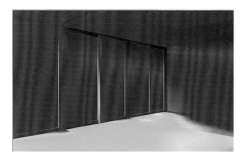
底层起居室南侧　South side, living room, 1F

底层餐厅西南侧　South side, dining room, 1F

建筑室内各部位冬季傍晚日照模拟计算结果
Simulation results of winter evening daylighting in various parts of the building

晴天时，建筑在夏季和春季的午间和傍晚以及冬季的午间都能得到非常出色的自然采光（春夏季的午间需要遮阳），室内光线最暗的部位也能达到阅读活动的照度标准。这个照度在以上季节的多云天气也能达到。左页图示显示的是冬季傍晚建筑室内各部位在晴天的情况下达到的照度，在 0.75m 的高度均达到了 150lx 以上照度，满足阅读的照度需求。在多云的春季傍晚室内也基本可以达到 150lx 的数值，仅在多云的冬季傍晚如在室内进行阅读活动需要人工照明辅助（此时即便是增加北侧的开窗面积也需要开灯补足）。

在模拟计算的同时，项目组还建了可拆解的实物模型，打算在同济大学光学实验室利用仿全天候自然采光装置进行两组不同立面的照度对比实验。这套模拟实验装置可以通过输入项目所在地的经纬度模拟太阳的实际运行轨迹，可惜无法记录建筑内的照度，也无法实现模拟计算，但是由于能够提供较为直观的观察效果，对设计学习者无疑是种体验。

On sunny days, the building receives excellent natural light during the midday and evening in summer/spring and midday in winter (shading needed in the midday of spring and summer), and the darkest area of the room can also meet the illuminance standards of reading activities. This illuminance can also be achieved in cloudy weather in the above seasons. The pictures on the left page shows the illuminance achieved in the various areas of the building in the winter on sunny days. At an altitude of 0.75 meters, the illuminance is 150 Lux, which satisfies the reading illuminance requirements. In the cloudy spring evening, the indoor illuminance can reach the value of 150 Lux. In the cloudy winter evening, it needs artificial lighting assistance for reading indoor, even by bigger window openings.

At the time of the simulation calculation, the project team also built a detachable physical model. It is planned to conduct illuminance comparison experiments on two different façades in the Optical Laboratory of Tongji University using the all-weather natural lighting device.

实物模型日照模拟实验
Physical model simulation experiment

能量系统
Energy System

3

要实现正能量,必须尽可能节省自身能耗和生产可再生能量。

To achieve energy plus performance it is necessary both to minimize energy consumption and to maximize renewable energy production.

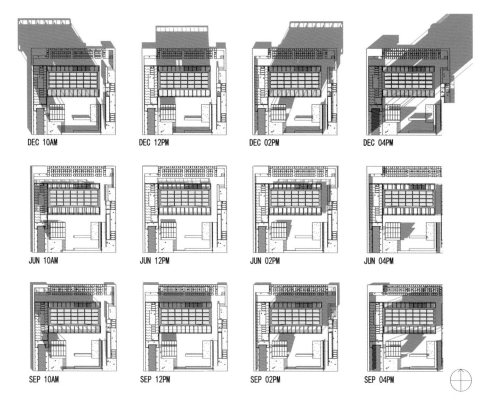

DEC 10AM   DEC 12PM   DEC 02PM   DEC 04PM

JUN 10AM   JUN 12PM   JUN 02PM   JUN 04PM

SEP 10AM   SEP 12PM   SEP 02PM   SEP 04PM

阴影图　SHADING

要实现正能量，必须尽可能节省自身能耗和生产可再生能量，具体的做法需要根据建筑所处的地理区位对温湿度、日照、风、水等自然环境进行分析后制定。"盒子"通过两个完整的系统来实现这个目标。

3.1 气候条件

我们要建造的参赛房屋位于德州市，是永久使用的建筑。德州市（37°26′2″N，116°21′26″E）属于寒冷气候区，大陆性气候，四季明确，夏季长、热、潮湿和部分多云，冬季很冷、干燥和大部分晴天：

- 在一年中，气温通常在 -5℃到 32℃之间变化，极少低于 -9℃或高于 37℃。年平均气温为 15℃；

To achieve energy plus performance it is necessary both to minimize energy consumption and to maximize renewable energy production. Doing this requires analyzing the natural environment — temperature and humidity, sunshine, wind and water, as determined by the location of the building. The "box" has two integral systems to achieve this goal.

3.1 Climate conditions

Dezhou City (37°26'2"N, 116°21'26"E) is in a cold climate zone with a continental climate and defined seasons. The summers are long, hot, humid and partly cloudy, while the winters are cold, dry and mostly sunny.

- During the year, the temperature usually varies between -5℃ and 32℃, rarely below -9℃ or above 37℃. The annual average temperature is 15℃;

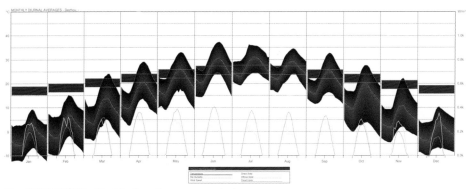

月、日平均气象值　Monthly diurnal averages, ecotect

Dezhou		Jan	Feb	Mar	Apr	May	Jun	Jul	Aug	Sep	Oct	Nov	Dec		
record high	℃	2.6	5.4	12.5	20.7	27.3	31.7	31.6	30.3	26.2	20.4	11.7	4.4	ø	18.73
record low	℃	-6.0	-3.7	2.2	9.4	15.5	20.5	23.3	22.5	16.7	10.2	2.9	-3.2	ø	9.19
daily mean	℃	-1.0	6.5	8.0	16.0	23.0	28.0	30.0	27.0	22.0	16.0	7.0	0.5	ø	15.25
average precipitation	mm	5.0	6.0	9.0	17.0	36.0	75.0	134.0	137.0	41.0	22.0	14.0	6.0	Σ	502.00
average precipitation days (≥0.1mm)	d	2.0	2.3	3.1	5.0	6.5	9.1	14.2	11.4	6.1	4.0	3.6	2.0	Σ	69.30
average relative humidity	%	57	56	54	49	49	56	76	79	69	66	70	70	ø	63
average sunshine hours	h/d	6.4	6.8	7.3	8.3	9.2	9.7	8.2	7.8	7.8	7.5	6.2	6.1	ø	7.61
sunny days	d	15.3	12.7	12.7	14.2	14.0	11.9	9.6	10.8	13.4	17.0	17.4	17.6	Σ	166.6
partly cloudy	d	9.1	8.9	11.8	10.9	11.6	13.9	15.9	13.5	10.2	7.7	7.0	7.7	Σ	128.2

德州气象数据　Climate data of Dezhou

- 年降水量约为 502mm，多集中于夏季，每年的变化很大；
- 一月是最凉爽、最干燥的月份，平均气温为 -1℃，平均降水量 5mm；
- 七月是最热和最潮湿的月份，相应的数字是 30℃ 和 134mm；
- 白天的长度在 9 到 15 个小时之间（最长在六月，最短在一月），日照量最多的月份七月里平均日照量将近 8 小时 / 天，最少十二月份也超过了日均 6 小时 / 天；
- 全年日照总量超过 2700 小时。因此德州市被誉为"太阳谷"。

- The annual precipitation is about 502mm, mostly concentrated in the summer, and the annual change is very large;
- January is the coolest and driest month, with an average temperature of -1℃ and an average precipitation of 5 mm;
- July is the hottest and wettest month, with an average temperature of 30℃ and an average precipitation 134mm;
- The length of the day is between 9 and 15 hours (the longest in June, the shortest in January), and July has the largest amount of sunshine, with an average of more than 8 hours/day. December has the least amount of sunshine; the average daily amount is 6 hours /day.
- The total amount of sunshine in the year exceeds 2,700 hours. Therefore, Dezhou City is known as the "Sun Valley".

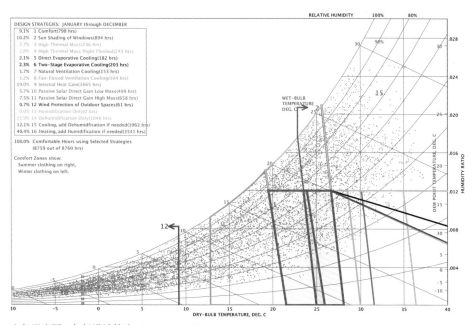

空气温度图，气候设计策略　Psychrometric chart, climate consultant

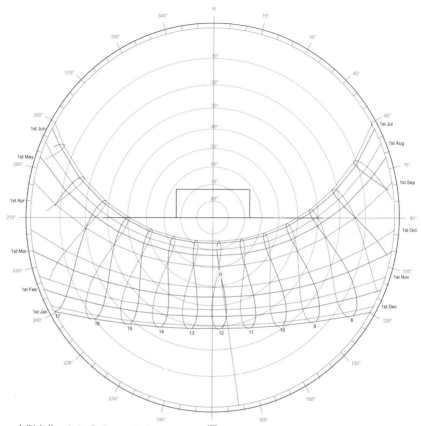

太阳方位　　Solar Paths, ecotect

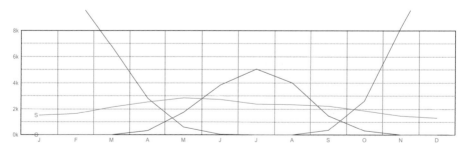

取暖月份Heating period:	制冷月份Cooling period:	舒适月份Comf. period:	太阳能获取Solar gain:
十月至四月Oct.-Apr.	五月至九月May-Sept.	五月初、九 Begin of May	最高: 五月High:May
供暖需求: 高	制冷需求: 中	月末、四月 End of Sept.&	最低: 十二月至一月
Heating demand: high	Heating demand:middle	Apr.	Low:Dec.-Jan.

度时数（取暖、制冷、太阳能）　DEGREE HOURS (Heating, Cooling and Solar)

Energy能量

- 取暖Heating
- 制冷Cooling
- 除湿Dehumidification
- 相变材料PCM

Facade立面

- 夏季最大得热来自东西立面（过热）；冬季最大得热来自南立面（用于取暖）
  Max. solar gaining in summer from the east/west facades in (overheating); the max. solar gaining in winter from the south facade (for heating)

- 夏季遮阳防护/冬季最大量获取太阳能
  Summer sun protection/winter max. solar gain

- 开窗面积：南侧70%，北侧40%，东西立面50-70%
  Window opening: 70% on south, 40% on north, 50-70% on east/west

- 应采用三层低辐射玻璃，Umax.=0.7W/m²K
  Triple low-e glass, Umax.=0.7W/m²K

- 阳光房和箱式窗
  Sun room and box window

- 最少200毫米厚保温或Umax.=0.12W/m²K
  Insulation min. 200mm or Umax.= 0.12W/m²K

Ventilation通风

- 年均1.7%可自然通风/适合热回收
  Average annual 1.7% natural ventilation/heat recovery reasonable

- 可利用早晚温差在某些月份进行夜间冷却
  The morning/evening temperature difference in some months allows night cooling

- 机械通风/热回收及除湿
  Mechanical ventilation/heat recovery & dehumid.

- 通风与烟囱效应结合
  Ventilation combined with chimney effect

Orientation朝向

- 夏季太阳入射角高/太阳能高；冬季太阳入射角低/太阳能低
  Summer high sun angle / large solar energy; Winter low sun angle / little solar energy

- 适合太阳能发电
  Suitable for solar power generation

- 适合使用太阳能光热设备-热水/取暖
  Suitable for solar thermal device - warm water/heating

- 水平面上太阳辐射最大
  Horizontal area receives max. solar radiation

- 中小进深有助于减少热损失/获得足够自然采光和通风；避免建筑过高以降低风压
  Medium/small depths help to reduce heat loss / obtain sufficient natural light/ventilation; avoid buildings too high to reduce wind pressure

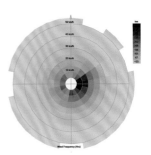

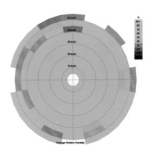

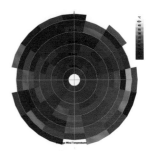

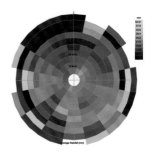

风玫瑰图　Windrose，ecotect

3.2 技术策略：主动式 + 被动式

3.2 Technical strategy: active + passive

• 被动式节能

在这样的气候特征条件下，为了满足舒适度要求，需要建筑室内夏季制冷、冬季采暖。其中由于建筑自身采用的被动式措施，冬季采暖系统的使用率较低甚至可以完全不用，这和普通建筑相比节省了大量的取暖能耗，因此是节能的。

• Passive energy-saving

Under such climatic conditions, cooling in summer and heating in winter are required. Due to the passive measures adopted by the building itself, the heating system for winter is rarely in charge or even completely unnecessary. This saves heating energy compared to an ordinary building.

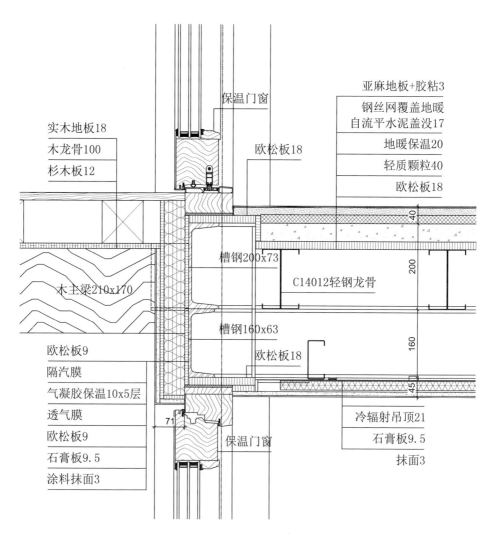

实木地板18
木龙骨100
杉木板12

保温门窗

欧松板18

亚麻地板+胶粘3
钢丝网覆盖地暖
自流平水泥盖没17
地暖保温20
轻质颗粒40
欧松板18

木主梁210x170

槽钢200x73

C14012轻钢龙骨

200

槽钢160x63

欧松板18

160

欧松板9
隔汽膜
气凝胶保温10x5层
透气膜
欧松板9
石膏板9.5
涂料抹面3

71

保温门窗

45

冷辐射吊顶21
石膏板9.5
抹面3

外围护结构：

第一个系统是它的被动式外围护结构系统，具体体现在：

- 轻质夹芯保温隔热外墙填充；
- 三层中空的保温窗（门）；
- 整体气密层；
- 支撑结构的断热处理；
- 加厚的屋面保温措施；
- 屋面除检修通道与排气孔处以外架空方式满铺太阳能光伏板，进一步增强屋面的保温隔热功能；
- 地面隔热处理；

Building envelope:
The first system is the passive building envelope:

- Lightweight sandwich insulation wall filling
- Triple glazing insulation window (door)
- Integral airtight layer
- Insulation of support structure
- Roof insulation measures
- The roof is covered with solar photovoltaic panels to further enhance thermal insulation
- Ground insulation treatment

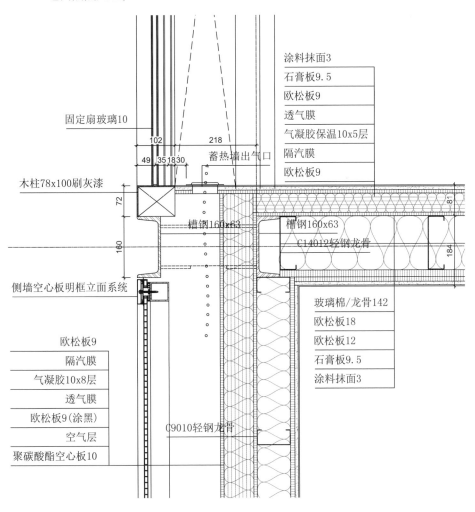

固定扇玻璃10

102   218
49 35 18 30
蓄热墙出气口

涂料抹面3
石膏板9.5
欧松板9
透气膜
气凝胶保温10x5层
隔汽膜
欧松板9

木柱78x100刷灰漆
72
160

槽钢160x63
槽钢160x63
C14012轻钢龙骨
81
184

侧墙空心板明框立面系统

玻璃棉/龙骨142
欧松板18
欧松板12
石膏板9.5
涂料抹面3

欧松板9
隔汽膜
气凝胶10x8层
透气膜
欧松板9（涂黑）
空气层
聚碳酸酯空心板10

C9010轻钢龙骨

- 南北缓冲区降低冬季外围护结构热传递损失；
- 南侧"阳光房"冬季为室内提供辅助热量；
- 南侧缓冲区为室内提供夏季自遮阳
- 东西侧墙采用空气墙构造增强轻质填充墙的保温隔热功能，降低通过热传递引起的能耗，并通过引入加热后的空气进入北部缓冲区，为其提供冬季热源；
- 南北通透，过渡季节进行自然通风。

- The north-south buffer zone reduces heat transfer loss in winter
- The southern "solar room" provides auxiliary heat for the interior in winter.
- The southern buffer zone provides self-shading for the interior in summer.
- The east and west side walls use solar wall construction to enhance the thermal insulation function of the lightweight sandwich wall. This reduces energy consumption caused by heat transfer and provides heated air for the northern buffer zone in winter.
- Natural ventilation during the transition season

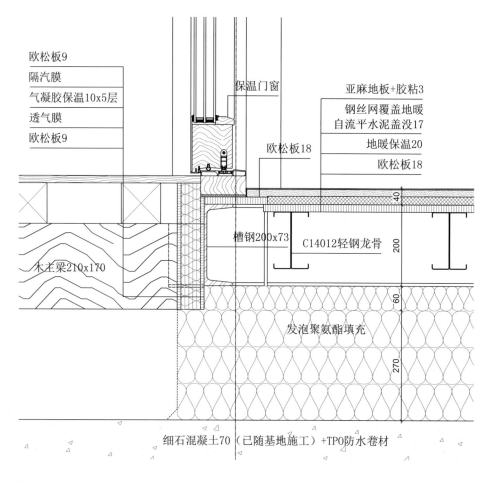

欧松板9
隔汽膜
气凝胶保温10x5层
透气膜
欧松板9

保温门窗

亚麻地板+胶粘3
钢丝网覆盖地暖
自流平水泥盖没17
地暖保温20
欧松板18

欧松板18

槽钢200x73   C14012轻钢龙骨

木主梁210x170

发泡聚氨酯填充

细石混凝土70（已随基地施工）+TPO防水卷材

- 太阳墙

正能量房 4.0 的另一个创新点是东西两面侧墙的太阳能光热利用，太阳墙的设计为多、高层建筑提供侧墙节能利用参考方案。

德州的地理位置偏北，通过模拟计算，在这个区位东西两侧墙安装太阳能光电装置受到照射时长和入射角度的限制，利用率较低，会影响建筑单位面积的产电效率，因此最终采用了开拓太阳能光热利用的方式，即根据太阳墙原理，用一块透明盖板封闭住外墙，盖板与外墙之间留有一定的间距，当阳光射入透明盖板，颜色较暗的实体墙面被晒热并加温与其直接接触的空气层，再引导被加热的空气进入室内或其他需要热空气的地方，利用由此获取的太阳能从而节能。对透明盖板的选择有一定的要求，最基本的要求是透明度高、隔热性好，此外还需要尽量轻质、耐老化、易清洁、低造价等。

东西侧墙上安装的太阳墙装置在冬季可作为室内的补充热源。太阳墙的空腔在二层楼板高度被分为两层，每层在底部和顶部各设置一个可开合的进出风口，冷空气从底部闸门进入空腔，经过太阳照射升温，变热的空气上升，再通过腔体上部预留的闸门进入需要供热的空间。这个空间可以按照实际使用需要来指定。由于北侧的缓冲区在冬季缺乏热源，这里基于竞赛展示的需要设置了一部楼梯，因此竞赛项目里将太阳墙的供热出口与北侧缓冲区相连。

- Solar wall

Another innovation of EnergyPLUS Home 4.0 is the solar thermal utilization of the sidewalls. The design of the solar wall provides a reference for energy saving utilization of side walls for multi-storey/high-rise buildings.

As a result of the simulation calculations, the solar PV panels on both sidewalls produce much less power than those on the roof. This affects the power production efficiency per unit area of the building. Solar heat usage is more suitable for these two walls, which are covered with a transparent material — leaving a gap between it and the walls. When the sunlight enters the cavity through the transparent cover, the air is warmed by the heated darker wall surface. The warm air rises and will be directed into the room or other places where heat is needed. There is an air inlet at the bottom and an air outlet at the top of the cavity. The cover needs to be transparent and have good insulation value. It should also be light, weather resistant, easy to clean, and inexpensive.

The cavity of the solar wall is divided into two parts at the position of the second floor. Since the buffer zone on the north side lacks heat supply source in the winter, the heat air outlet of the solar wall is connected to the north buffer zone.

西立面，聚碳酸酯板分隔图，无比例　　West Elevation, polycarbonate plate layout, no scale

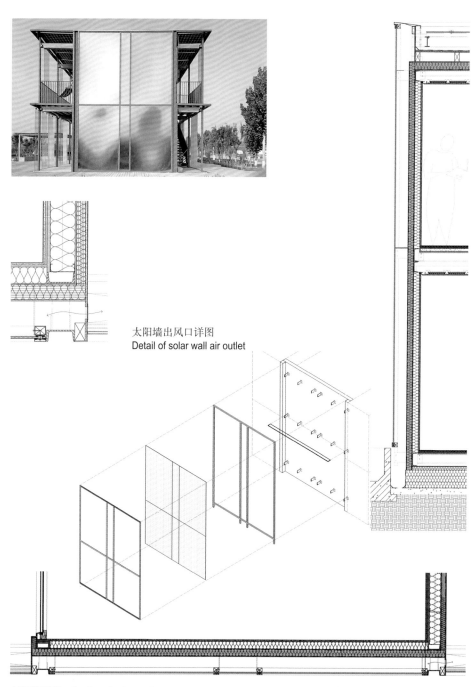

太阳墙出风口详图
Detail of solar wall air outlet

太阳墙剖面　Section of solar wall

● 制冷和供暖系统

正能量房 4.0 项目中采用了将制冷和取暖分开并分别置入天花板内和地板内的两套系统，通过水循环制冷和取暖。这两套系统都是同样的热（冷）辐射空调系统，只用一套安装在吊顶里的系统即可满足夏季制冷和冬季取暖的功能。两套系统同时安装的主要目的是提高舒适度——夏季冷辐射来自天花板，冷量下行，可以更均匀地分布在室内；冬季热辐射来自于地面，体感更为舒适。另一个目的则是充分展示"盒子"休系各部分空间的可利用能力——用户可以根据各自的喜好选择将空调系统安装于天花板内还是地板内；第三层原因是希望在过程中实测不同产品的工作性能——安装在天花板内的水循环冷辐射系统采用的管径较粗、管间距较大，而地板内的热辐射取暖系统则采用了毛细管垫。

• Cooling and Heating System

The EnergyPLUS Home 4.0 project uses two water cycle systems for cooling and heating, installed in the ceiling and the floor. Both systems are thermal (cold) radiant air conditioning systems that can be used for summer and winter. The main purpose of using two systems is to improve user comfort: Cold radiation from the ceiling in the summer can be more evenly distributed in the rooms; heat radiation from the floor in winter is more comfortable. Another purpose is to demonstrate the availability of the various parts of the box — users can choose whether to install the air conditioning system in the ceiling or the floor. In addition, it is possible to measure the performance of products in the project. The pipe diameter and spacing of the radiation system in the ceiling is larger than that of the heating system in the floor.

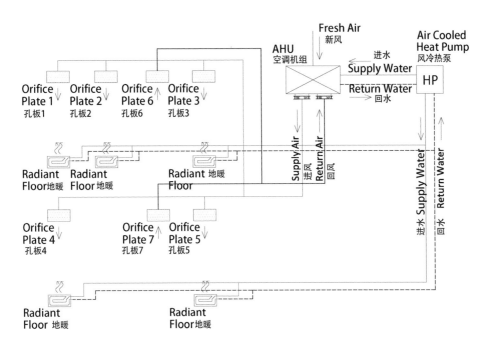

建筑室内空气环境控制系统图　Indoor environment control system diagram

• 新风系统

被动式的外围护系统必须配一套新风系统。在每个"盒子"的顶部结构梁的高度范围内，都根据功能分区布置了风管。主管道位于底层起居室和餐厅的北侧，分别汇入服务功能盒子单元内的主风管，再进入位于设备间顶部的室内风机；一个竖向的管道将新风同时送到二层，在二层风管同样位于北侧的交通空间吊顶内。由于风管的主要功能是为室内提供新鲜空气，制冷和供暖只是辅助性的，所以风管的管道截面较小，可以充分利用梁高，在承重结构允许的工字梁腹板部位留出风管刚好可以穿过的洞口。

室内风机位于底层的设备间，室外进风管道位于底层盒子下面的架空层里（底部保温层），从设备间底部进入室内管井并与室内风机相连。

• Fresh air system

Passive enclosure systems must be equipped with a fresh air system. The air duct is located in the north side of both floors in the top structural beam of each box. The main ducts enter the service box unit and join the indoor fan hanging on the ceiling of the equipment room; a vertical pipe sends the fresh air to the second floor. Since the main function of the system is to provide fresh air to the rooms and the cooling and heating functions are only secondary, the duct section is very small and can be embedded in the beam.

The indoor fan is located in the equipment room on the first floor. The outdoor air inlet duct enters the indoor duct well from the bottom of the box and is connected with the indoor fan.

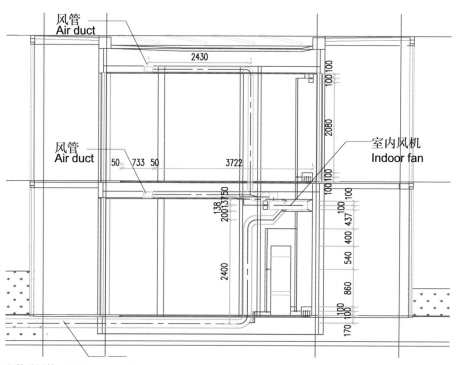

室外进风管　Outdoor air duct

• 主动式产能

和被动式建筑相比，主动式建筑由于更关注建筑的产能性能因而无需完全达到被动式建筑对建筑外围护结构保温与气密性的严格要求，这样，建筑在表现和使用上都获得了更多的灵活空间。

产能最大化原则：

尽可能利用建筑自身、基地内部和场地周边所有可利用的条件生产可再生能量。由于历史性原因，太阳能竞赛规则不允许使用地热，而建筑本身较为低矮，根据基地的风能分析，也不适合主动式利用风能，因此在正能量房4.0项目中，唯一可直接利用的可再生能源是太阳能，这也是德州地区气候特征中的优势。

能效优化原则：

原则上来说，建筑的屋顶、南立面、东西立面都应该考虑尽可能用做产能，某些纬度适合地区甚至北立面也会接收相当多的太阳能辐射。但是这些

• Active energy production

Compared to the passive buildings, active buildings do not need to fully meet the strict requirements of passive buildings for insulation and airtightness. The focus is on the energy production capacity of the building, which allows for flexibility in form and function.

Energy production maximization:

Using all available conditions within the building, the site, and area around the site to produce as much renewable renergy as possible. The competition does not allow the use of geothermal energy, and wind energy is not suitable for such low buildings, so the only renewable energy source is the sun—an advantage given by the local climate conditions.

Energy efficiency optimization:

In principle, the roof, south façade, and east and west facades of the building should be considered for energy production. However, these façades have several other functions, such

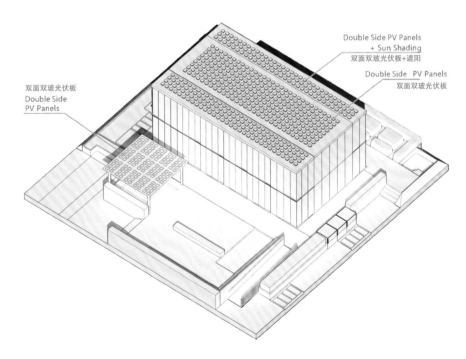

立面还有若干使用功能需要满足，如采光、通风、视线要求等。根据比赛规则，单位面积上所安装的太阳能光伏板的产能效率占有能源项目的 20% 分数，因此必须综合考虑上述立面的功能以及单位面积上的年光辐射总量。根据模拟计算，德州地区由于纬度不算太高，立面所获得的太阳能辐射量远低于屋面，根据建筑的形体和空间特征，结合竞赛的规则要求，正能量房 4.0 的能量生产策略定为集中考虑建筑和场地可以利用的屋顶面积，而立面则专注于被动式节能。

产能模拟计算：
将生成的建筑模型导入软件（Autodesk Ecotect Analysis），设定建筑所在地域，分别针对屋顶和立面进行年产能计算。在忽略太阳能板的反射，计算的是在效率 100% 的情况下，安装在屋顶的太阳能板年产电量为 3204Wh/m²，南立面上的太阳能板年产电量为 2502Wh/m²，相对屋顶位置的太阳能板低 28%。

as lighting, ventilation, and view requirements. According to the competition rules, the energy production efficiency of photovoltaic panels installed per unit area occupies 20% of the energy performance category. Therefore, it is necessary to consider both the facade function and the energy production efficiency factor. The simulation calculation shows solar radiation obtained from the façade is much lower than that from the roof. According to the form and spatial characteristics of the building and the competition rules, the energy production strategy of EnergyPLUS Home 4.0 focuses on the roof area of the building and the site, while the facade focuses on passive energy-saving.

Energy production simulation:
Use the software (Autodesk Ecotect Analysis) and import the building model, set the building location, and calculates the annual capacity for the roof and facade respectively. By neglecting the reflection of the solar panel and 100% working efficiency, the annual output of the solar panels installed on the roof is 3204 Wh/m2, and on the south facade is 2502 Wh/m2, which is 28% lower to the solar panels on the roof.

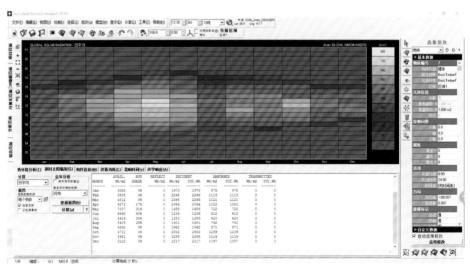

建筑产能模拟计算　Energy production simulation

材　料
Materials

4

正能量房 4.0 的设计理念来自德国，但是由于房屋的建造地点在中国德州，使用者和面对的参观者大多数来自国内，所以我们必须从本土市场出发，将实现项目的关键环节都建立在中国市场的可行性上，所选用的建筑产品、建筑材料和建筑商均以国内企业为主，来自国外的供应商也需要有国内的合作关系或意向。

The architectural concept of EnergyPLUS Home 4.0 comes from Germany. But the house was built in Dezhou, and most of the users and visitors are from China. So we must focus on the local market. The building products, materials, and constructors are mainly from China.

4.1 外围护结构——透明构件

正能量房 4.0 的设计理念中至关重要的是外围护结构中的透明部件，它们面积大，保温性能、透光性、构件耐久度等要求严格，产品质量能否满足设计要求决定了项目的成败，因此在项目深化阶段最先要确定的就是门窗产品。为此我们进行了广泛的国内外市场调研，经过多次考察和论证，最终选定了米兰之窗（www.miluxwindows.com）作为门窗的供应商，而米兰之窗也对项目给予了最大的支持，并且与学院建立了战略合作伙伴关系。

项目中最大的窗洞位于底层南侧，净高 2680mm，净宽 2280mm。这里全部为推拉落地窗，即单扇窗的净尺寸为 2680mm×1140mm。相邻两樘的推拉扇中轴对称，当全部窗扇都打开时，两樘窗之间形成一个开间的开启，以实现最大的室内外一体。这扇大尺寸的窗有三层玻璃，其窗框不仅要能承受自身的重量，还要符合瑞士低能耗建筑对门窗保温的要求。当然，由于这是推拉窗，其开启和移动必须是灵活、轻便的。

4.1 Building envelope: Transparent components

To realize the design concept of EnergyPLUS Home 4.0, it is essential to ensure the performance of the transparent building components. They have a large area, strict requirements such as thermal insulation performance, light transmission, and durability. Whether product quality can meet the design requirements determines the success or failure of the project. After many investigations, we selected windows by Milux (www.miluxwindows.com), which sponsored insulation windows and doors for the project.

The largest window opening on the south side of the first floor has a net size of 2680mm×2280mm. These are triple-glazed sliding windows with the floor height. The single window size is 2680mm×1140mm. Though the windows are heavy, they should be easily moved and slide smoothly.

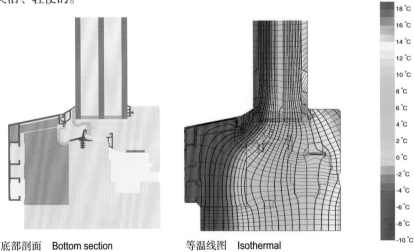

底部剖面　Bottom section　　　　等温线图　Isothermal

测试窗数据：木框［0.13W/(m² · K)］外包铝合金，带 PU 保温层［0.036W/(mK)］。窗玻璃为三玻两腔：47 mm(5/16/5/16/5)

Testing window data: Timber frame［0.13W/(mK)］with external aluminium shall and insulation PU［0.036W/(m² · K)］. Used Pane: 47 mm (5/16/5/16/5)

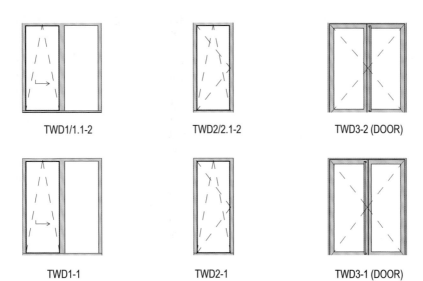

TWD1/1.1-2   TWD2/2.1-2   TWD3-2 (DOOR)

TWD1-1   TWD2-1   TWD3-1 (DOOR)

WINDOW SCHEDULE			
TYPE 类型	LEVEL 位置	WIDTH (mm) 宽度	HEIGHT (mm) 高度
TW2-1	Ground floor 底层	1155	2685
TW1-1	Ground floor 底层	2280	2685
TW1-1	Ground floor 底层	2280	2685
TW1-1	Ground floor 底层	2280	2685
TW1-1	Ground floor 底层	2280	2685
TW1-1	Ground floor 底层	2280	2685
TW2-1	Ground floor 底层	1155	2685
TW2-1	Ground floor 底层	1155	2685
TW2-1	Ground floor 底层	1155	2685
TW2-1	Ground floor 底层	1155	2685
TW2-2	Upper floor 二层	1155	2485
TW2-2	Upper floor 二层	1155	2485
TW2.1-2	Upper floor 二层	1155	2485
TW2.1-2	Upper floor 二层	1155	2485
TW1-2	Upper floor 二层	2280	2485
TW1-2	Upper floor 二层	2280	2485
TW1-2	Upper floor 二层	2280	2485
TW1-2	Upper floor 二层	2280	2485
TW1-2	Upper floor 二层	2280	2485
TW1.1-2	Upper floor 二层	2280	2485
TW2-2	Upper floor 二层	1155	2485

	$U_{f\text{-value}}$ [W/(m²K)]	W_{idth} [mm]	ψ_g [W/(mK)]	$f_{Rsi=0.25}$ [-]
Spacer			SWISSP. Ultimate*	
Bottom Fix	0.84	116	0.028	
Bottom S	0.90	116	0.028	
Top Fix	0.78	93	0.029	
Top S	0.88	93	0.029	0.71
Side Fix	0.76	93	0.028	
Side S	0.82	93	0.028	
Mullion	0.90	133	0.028	

竖梃　Mullion

窗框热工数据　Thermal data for the window frame

以上这些条件是对国内门窗制造商提出的极为严格的要求。作为行业中的领军企业，米兰之窗接受了挑战。在项目设计阶段，米兰之窗已经研制出了保温推拉窗，并获得了德国被动式建筑研究所的节能证书。与不具备保温性能的传统落地式推拉门窗不同，保温推拉门窗在关闭状态下移动扇与固定扇呈企口咬合状并通过压紧的胶条满足气密性要求。关紧的门窗处于同一个平面，这样在保温节能的同时也改善了室内效果，可谓一举两得。

The Milux's products meet those demanding requirements. At the project design stage, Milux has developed the insulation sliding window and obtained an energy-saving certificate from the German Passive Building Institute. When the sliding window is closed, the moving sash and the fixed sash are in the same plane, and the window meets the airtightness requirement by pressing the rubber strip tightly. This improves the visual effect and reduces the heat loss.

南侧的双扇推拉落地窗　Sliding windows/doors on the south side

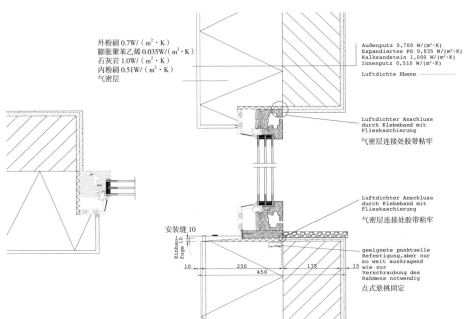

外粉刷 0.7W/（m²·K）
膨胀聚苯乙烯 0.035W/（m²·K）
石灰岩 1.0W/（m²·K）
内粉刷 0.51W/（m²·K）
气密层

Außenputz 0,700 W/(m²·K)
Expandiertes PS 0,035 W/(m²·K)
Kalksandstein 1,000 W/(m²·K)
Innenputz 0,510 W/(m²·K)

Luftdichte Ebene

Luftdichter Anschluss
durch Klebeband mit
Flieskaschierung
气密层连接处胶带粘牢

Luftdichter Anschluss
durch Klebeband mit
Flieskaschierung
气密层连接处胶带粘牢

安装缝 10
Einbau-fuge 10

10   250   175   15
450

geeignete punktuelle
Befestigung,aber nur
so weit auskragend
wie zur
Verschraubung des
Rahmens notwendig
点式悬挑固定

安装要求　Installation requirments

北侧的单扇平开内倒落地窗（门）　Swing/tilt windows/doors on the north side

4.2 外围护结构——不透明构件

与透明的外围护结构构件相比，不透明的墙体适用材料涉及种类较多，可选范围较大，但是真正能够符合正能量房4.0项目的要求却不那么容易做到。

由于建筑是采用钢框架体系，并且需要遵循装配式原则，墙体材料必须满足轻、薄、模数化要求，要有高效的保温性能，同时根据建筑规范，尽量做到不燃、耐久。经过长时间的材料研究和市场调查，只有真空隔热板［导热系数0.008W/(m²·K)］和玻璃纤维气凝胶［0.013～0.02W/(m²·K)］可以满足设计要求。这两种材料都是不燃材料，在我国均有生产。另外还有一种只有德国赢创才生产的材料CALOSTAT［导热系数0.019W/(m²·K)］，也是可以满足设计要求的不燃保温材料，但是价格昂贵，并且目前仅供应德国和欧洲市场，初步接触后即排除了使用的可能性。

4.2 Building envelope: Opaque components

Although there are more opaque wall materials than transparent wall materials, it is still not so easy to meet the requirements of the EnergyPLUS Home 4.0 project.

Since the building adopts the steel frame system and needs to follow the assembly principle, the wall material must be light, thin and modular, and it must have high insulation value. According to the building code, it should be as non-combustible and durable as possible. Our material and market research has found that only vacuum insulation panels ［λ=0.008 W / (m² · K)］ and the glass fiber aerogel ［λ=0.013-0.02 W / (m² · K)］ can meet the design requirements. Another expensive material CALOSTAT ［λ=0.019W/(m² · K)］ can also meet the design requirements. But it is only produced by Evonik in Germany and not available in China.

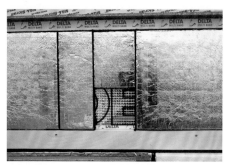

2007 年美国太阳能国际竞赛德国队项目外墙保温预制施工过程
The prefabrication of the exterior wall insulation using VIP, Team Germany of SDC 2007 in US

真空隔热板（VIP）使用多孔的高温蒸馏的原硅酸质芯材与高阻气复合薄膜，通过抽真空封装技术制成。在 2007 年美国举办的太阳能十项全能竞赛中获得冠军的德国队项目使用了这种保温材料，保温层仅有 6cm 厚（2cm×3cmVIP 板），可以达到的 U 值为 0.1W/（m²·K）[详见《主动式建筑：从被动式建筑到正能效房：*Vom Passivhaus Zum Energieplushaus*》（德）曼弗雷德·黑格尔（Manfred Hegger）编著，曲翠松译]。因此，在项目深化设计的初始阶段，项目组也是以真空隔热板作为主要目标材料进行考察的。但是经过反复接触和实地调研，发现正能量房 4.0 实际上无法使用这种材料。

首先由于建筑是采用钢结构承重，露在外面的需要特别保温的钢构件断面极小，而真空隔热板的加工尺寸必须满足一定的长度和宽度，太小了无法加工。这实际上已经一票否决了使用真空隔热

Vacuum insulation board (VIP) is a high-efficiency insulation board made from porous, high-temperature distilled orthosilicate core material sealed by a high-airtight composite film with vacuum encapsulation technology. This was used by the 2007 SDC winner project. The insulation layer was only 6 cm thick (2cm×3 cm VIP board) with a U-Value 0.1 W/(m²K). (See *Vom Passivhaus Zum Energieplushaus*, by Manfred Hegger, translated into Chinese by Cuisong Qu). In the initial developing design stage, the project team investigated vacuum insulation board as the target material. However, after repeated contacts and field visits, we found out that the project could not use this material.

First, the processing dimensions of the vacuum insulation board must meet a certain length and width, and the steel components that need to be insulated are extremely small. Secondly,

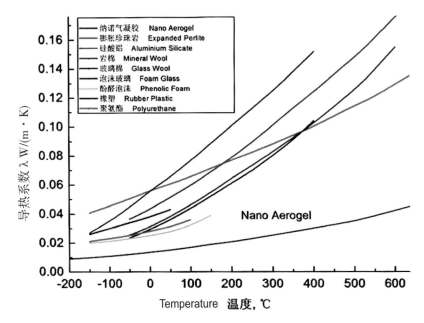

不同温度下气凝胶与传统材料导热系数对比图
Comparison of thermal conductivity between aerogel & conven materials at different temperatures

板做外保温材料的可能性。其次，小尺度构件外使用真空隔热板施工难度很大，基本上处理不了接缝处保温性能差的问题，结果势必会造成大量的冷桥；再者，高造价也是一个极大的阻碍。另外，材料还存在真空度难以保持的问题，高阻气膜随着时间推移总会漏气，其保温性能将会大大降低乃至完全失效。

项目组最终选择了使用玻璃纤维气凝胶。它的实验室测试数据虽然不如真空隔热板好，但却拥有很多其他优点。气凝胶材料以纳米二氧化硅气凝胶为主体材料，通过特殊的工艺复合而成，具有耐高温、导热系数低、密度小、强度高、绿色环保、防水不燃等优越性能，同时兼具优越的隔声减震性能。玻璃纤维气凝胶则是用玻璃纤维高强度材料作为气凝胶的承载体，赋予其相应形状，以便加工使用。纳诺科技有限公司（www.nanuo.cn）生产的玻璃纤维气凝胶绝热毡为厚度 6mm 或 10mm 的卷材，这种产品集保温性能、防火性能于一体，易于切割加工，是很难得的材料，但是目前还没有大面积地应用于建筑工程领域，主要原因是这种产品的造价太贵，用于工业管道及交通领域时性价比很高，但是建筑保温材料的用量较大，在建筑节能法规要求不高时可选的材料较多，所以除了几个示范项目还没有工程案例将它用于外墙外保温，其可靠性还有待验证。

基于以上综合性原因，加上纳诺科技有限公司与同济大学物理实验室的长期合作，正能量房 4.0 的整体外墙保温材料最终选择了玻璃纤维气凝胶，并由

the construction of vacuum insulation panels on small-scale components is very difficult, and the problem of poor insulation performance at the joints cannot be solved, which will inevitably result in a large number of cold bridges. Third, the vacuum is difficult to maintain, and the film will always leak, Its thermal insulation performance will be greatly reduced or completely fail over time.

The project team finally chose to use glass fiber aerogel, although the laboratory test data are not as good as those for vacuum insulation board, Still, there are many other advantages. The aerogel material is made of nano-silica aerogel by a special process. It has high-temperature resistance, low thermal conductivity, low density, high strength and is green. It is waterproof, non-combustible, and has superior sound insulation and shock absorption performance. Glass fiber aerogels use glass fiber as carriers for aerogels. The glass fiber aerogel insulation felt produced by Nano Technology Co., Ltd. (www.nanuo. cn) is a coil of 6 mm or 10 mm thickness. This thermal insulation product is fireproof, easy to cut and process, and used widely in industrial pipeline and transportation instrument insulation. Due to its high price, this material has been rarely used in construction. Except for a few demonstration projects, there are no building references for exterior insulation, and its reliability has yet to be verified.

There is a long-term cooperation between Tongji University and Nano Technology Co., Ltd., which sponsored the glass fiber aerogel

纳诺科技有限公司气凝胶保温产品，来源：纳诺科技
Aerogel insulation products of Nano Technology Co., Ltd. source: Nano Technology

纳诺科技有限公司全部赞助。钢结构外侧和轻质外墙的支撑及填充外侧均使用5层厚度为10mm的玻璃纤维气凝胶绝热毡，层与层搭接处进行错缝处理，提高保温度。墙体的计算 U 值达到 0.26，这虽然与瑞士低能耗建筑的外墙保温技术要求相差近一倍，但是大大高于我国节能建筑法规要求。这个指标是同时考虑了各方面因素后最终确定的，设想通过项目建成后的实际使用效果验证指标的合理性。

products for the project. The steel structure and the lightweight walls are insulated with five layers of glass fiber aerogel insulation felt, and the total thickness of the insulation is 10 mm. The felts are overlapped to improve the insulation effect. The calculated U-Value of the wall reaches 0.26. This is nearly double the technical requirements of the external wall insulation of the Swiss Minergie buildings, but much lower than the China's energy-saving building regulations required.

上左：运至预制工场的卷材气凝胶隔热毡
上右：预制安装侧墙气凝胶保温施工现场
左：使用保温钉固定
下：断面较细的钢梁祝保温层需现场施工
Upper left: Aerogel insulation felt coils
Upper right: Construction of sidewall insulation
Left: Fixed with insulation nails
Bottom: Construction of steel beam/column insulation on-site

4.3 缓冲区——透明构件

阳光房作为缓冲区，首先要满足功能上的需要，即在冬季保证光线最大限度射入，而变暖的空气既不会从缝隙渗漏出去，也不会通过热传导而流失。这实际上要求它的外围护结构透光率高、导热性差、密封性好。其次，阳光房在夏季应尽可能通透，以免热量积聚，产生温室效应，使得与之相邻的室内温度受到影响。这就要求开启扇面积大，开启扇开启后所占空间较小，以免影响缓冲区的使用。此外，缓冲区位于主体建筑之外并将其围裹，这层结构也就是外立面，其外观效果极为重要。

从概念设计方案中遴选出的方案中，缓冲区外立面分为上下两层，在屋顶、楼板、地面之间是均质的竖向透光板。这个极简的风格非常重要，在深化和实施过程中需尽量保留。这样，位于二层的透光板高度是上层盒子高度再加上屋顶女儿墙的高度，达到3.20m，底层稍短，为2.9m。每个盒子对应的开间宽度四等分，靠近平台立柱的左侧或右侧设一扇固定透光板，其余三扇均为横向推拉扇。这样在夏季的炎热天气里，缓冲区呈打开状，三扇推拉扇可以平行排布隐藏在固定扇之后，实现75%的最大开启面积。

原方案设想采用常用的玻璃板来做透光板。玻璃是成熟的工业建材，在强度允许的情况下可以做成各种尺寸，易清洁、耐老化，但是比重大，易碎，对支撑构

4.3 Buffer zone: Transparent components

The buffer zone is conceived as a solar room and necessary to ensure maximum light penetration during the winter. The warm air should not leak from the gaps or through heat conduction. This requires its enclosure structure to have high light-transmittance, poor thermal conductivity, and good sealing. Secondly, the solar room needs to be opened as much as possible during the summer to avoid the greenhouse effect.

The buffer zone is divided into upper and lower parts and enclosed by homogeneous vertical light-transmissive plates of floor height. This minimalist style is important for the architecture and needs to be preserved as much as possible in the design development and implementing. Thus, the light-transmissive plates on the second floor are the height of the upper box plus the roof height, 3.20 meters in total. On the first floor, the height is 2.9 meters.

The light-transmissive plates are supposed to be ordinary glass plates in the original plan. Glass is a mature industrial building material. It can be cut to size, is easy to clean and resists aging. But it is heavy, brittle, and requires stronger support members. The steel supporting components of the buffer zone are designed to be as thin as possible to achieve a light and transparent effect. Thus we decided to try an innovative synthetic polycarbonate

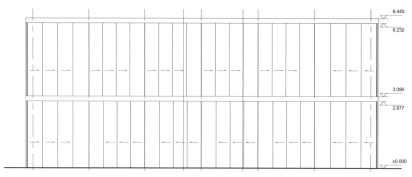

南侧缓冲区立面   Elevation of southern buffer zone

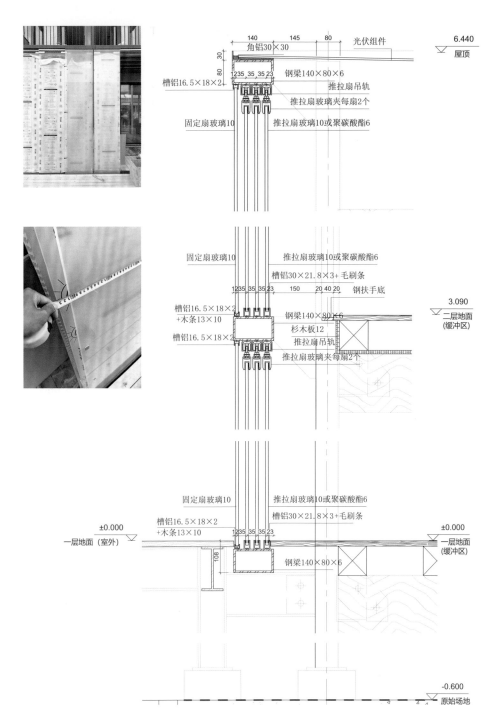

140　145　80　光伏组件

角铝30×30

30

80

槽铝16.5×18×2

1235,35,35,23

钢梁140×80×6

推拉扇吊轨

推拉扇玻璃夹每扇2个

固定扇玻璃10

推拉扇玻璃10或聚碳酸酯6

6.440

▽

屋顶

固定扇玻璃10

推拉扇玻璃10或聚碳酸酯6

槽铝30×21.8×3+毛刷条

1235,35,35,23　150　20,40,20　钢扶手底

槽铝16.5×18×2
+木条13×10

槽铝16.5×18×2

钢梁140×80×6

杉木板12

推拉扇吊轨

推拉扇玻璃夹每扇2个

3.090

▽

二层地面
(缓冲区)

固定扇玻璃10

推拉扇玻璃10或聚碳酸酯6

槽铝30×21.8×3+毛刷条

±0.000

▽

一层地面 (室外)

槽铝16.5×18×2
+木条13×10

1235,35,35,23

108

钢梁140×80×6

±0.000

▽

一层地面
(缓冲区)

-0.600

▽

原始场地

53

件要求较高。缓冲区的钢构件追求尽量纤细，以达到轻盈通透的效果，因此在深化设计时综合考虑上述多方面要求，决定尝试部分采用一种创新型合成材料聚碳酸酯板（实芯板材，科思创公司德国工厂生产，上海公司提供产品赞助与技术支持，www.covestro.com）。

聚碳酸酯板最大的优点是轻质、透明、不易碎、可塑性强。与玻璃相比，它的另一个更适合本项目的特征是热传导系数较低〔0.19 W/（m²·K），玻璃为1.11W/（m²·K）〕。但由于优质聚碳酸酯板造价昂贵，项目最终选择了最能体现材料特性的部位——南侧缓冲区的推拉扇使用这种材料，北侧缓冲区不具备阳光房的功能，仍然保留原方案设计使用传统玻璃。所有固定扇采用磨砂玻璃，当缓冲区完全开启时，在建筑正面只看到这些比例细长的磨砂玻璃固定扇。

能够实现阳光房这个概念，并且给予建筑一个令人耳目一新的外观，其中功能性五金构件功不可没。海迪克五金（www.hideca.com.cn）赞助了推拉扇要完成顺滑推拉的全部五金件。每个推拉扇顶部都有两个装有滑轮的不锈钢夹，其夹口宽度可调整，前后缓冲区正面和侧面加起来共有88个这样的推拉扇，也就是说项目一共需要176个滑轮钢夹，分别将玻璃板（10mm厚）或聚碳酸酯板（6mm厚）固定住顶端再吊挂在安装于钢横梁底部的铝合金导轨上（均由海迪克建筑材料有限公司免费提供）。

plate, produced by Covestro in Germany (Covestro Shanghai sponsored the products and provided the technical support. www.covestro.com).

It is light, transparent, non-fragile, plastic and it has a much lower heat transfer coefficient [λ=0.19 W/m•K, 1.11W/(m²•K) by glass]. However, due to the cost of high-quality polycarbonate plates, the project chose to use this material for the sliding leaves in the south side buffer zone. The north side buffer zone does not have the function of the solar room and uses ordinary glass. All fixed parts use the frosted glass. When the buffer zones are fully open, only those fixed slim frosted glass plates are visible on the front of the building.

There are two stainless steel clips with pulleys on the top of each sliding element. The width of the jaws can be adjusted. There are 88 such sliding elements in the front and back buffer zones, which makes a total of 176 steel clips with pulleys. The glass plate (10mm thick) and polycarbonate plate (6mm thick) hang on the aluminum alloy rail mounted on the bottom of the steel beam though those steel clips. The realization of the concept of the solar room also depends on the functional hardware components. Hideck Hardware (www.hideca.com.cn)sponsored all the hardware for the sliding elements.

海迪克建材功能性五金
The functional hardwares of Hideck Hardware

不锈钢玻璃夹、滑轮，铝合金吊轨
Stainless steel clips, pulleys and aluminum rails

4.4 防水系统

　　建筑防水的重要性在这里不做赘述，与本项目的合作伙伴东方雨虹（www.yuhong.com.cn）专业的技术团队一起，我们寻求的是系统性的解决方案，要能够满足高效率节能、便捷施工和隔热（湿）和密封性的一体化要求。

　　建筑系统性防水方案包括地基防水、墙面防水和屋面防水。东方雨虹在方案深化阶段提供了多种防水材料可供选择，从改性沥青防水卷材，到合成高分子卷材以及方式砂浆等，考虑到展示性和施工便捷性，项目最终选用了柔性防水解决方案，使用雨虹 TPO 高分子防水卷材（基础1.2mm 厚、屋面 1.5mm 厚），隔热隔汽材料为杜邦特卫强（Tyvek）8314X 反射性隔汽膜（采用双面丁基胶带进行点粘结的方式基面铺贴施工处理）。

　　TPO 防水卷材是采用先进聚合技术将乙丙（PE）橡胶与聚丙烯结合在一起的热塑性聚烯烃（TPO）材料为基料、以聚酯纤维网格织物做胎体增强材料构成，并采用先进加工工艺制成的可卷曲的防水材

4.4 Waterproof System

Together with the professional technical team of the project's partner, Oriental Yuhong (www.yuhong.com.cn), we worked for a solution to meet high efficiency, energy-saving, convenient construction, heat and wet insulation sealing requirements.

Waterproofing system solutions include foundation waterproofing, wall waterproofing and roofing waterproofing. Oriental Yuhong offers a variety of materials, from modified asphalt waterproofing membranes to synthetic polymer coils and mortars. The project chose a flexible waterproof solution, using Yuhong TPO polymer waterproof membrane (1.2 mm thick by foundation, 1.5 mm thick on the roof) and DuPont Tyvek® 8314X reflective vapor barrier film (Base surface paving construction treatment by double-sided butyl tape for point bonding).

TPO waterproofing membrane has a white, smooth surface, which is highly reflective

太阳能竞赛施工现场：TPO 防水卷材裁切
SDC construction site: TPO waterproof coil cutting

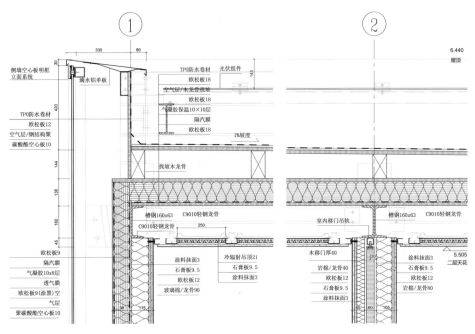

防水层结构详图：屋面、墙屋面
Waterproof construction details: roof, wall and roof

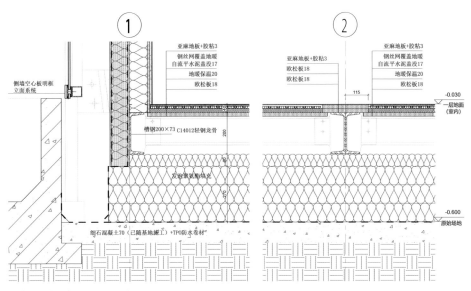

防水层结构详图：地面、墙地面
Waterproof construction details: ground, wall and ground

料。它兼有乙丙橡胶优异的耐候性和耐久性与聚丙烯的可焊接性，搭接采用热焊接，可形成高强度的密封层，使得施工操作快速便捷。

其次，材料采用特殊的配方技术，具有高柔韧性，不会产生一般聚烯烃材料因增塑剂迁移而变脆的现象；其优异的耐高低温柔韧性能可在 –50℃下仍保持柔韧性，在较高温度下保持机械强度，保持长久的防水功能。

TPO 防水卷材为以白色为主的浅色，表面光滑，高反射率，具有较好的节能效果；且耐污染。成分中不含氯化聚合物或氯气，焊接和使用过程中无氯气释放，对环境和人体健康无害。在现场进行裁剪和焊接产品，能够满足各种型号和尺寸要求。

and has good energy-saving capability and pollution resistance. The composition does not contain chlorinated polymer or chlorine gas. That means there is no danger from chlorine being released during welding and use. So the material can be cut and welded on site.

屋面 TPO 防水
TPO waterproof on the roof

基础 TPO 防水
TPO waterproof of the foundation

太阳能竞赛施工现场：TPO 防水卷材焊接
SDC construction site: TPO waterproof coil welding

4.5 地板——福尔波亚麻地板

正能量房 4.0 的冬季采暖措施里采用毛细管网低温水循环进行热辐射供暖，这种方式不仅节能，还能提供最大限度的热舒适度。为了保证毛细管网地板热辐射的工作效率，必须选择合适的地板材料。

为了使室内能够快速升温，毛细管网以上的面层厚度越薄越好。作为楼地板面层，材料必须有足够的强度，耐磨，易清洁。此外，项目本身的特殊性要求铺装方便、达到较好的视觉效果。以上这些要求福尔波亚麻地板都能做到。

亚麻地板历史悠久，至今已经超过了 150 年，是弹性地材行业的开山鼻祖。它的原型是一种"蜡"或"油布"，这种材料是在 1627 年被发明的，在经过了 200 多年以后，才逐渐被发展为地板。苏格兰人 Walton 先生于 1860 年发明了一种大量生产亚麻籽油覆层的方法，并将其命名为 Linoleum（在拉丁语中，Linum 意为亚麻，而 oleum 意为油），可认为是亚麻地板正式诞生。正能量房 4.0 上下两层一共 150m² 的室内地面全部采用福尔波集团（www.forbo.com）生产的梦梵丽亚麻环保地板（Forbo Marmoleum）。

注册于苏黎世，主要生产机构位于荷兰克罗默尼的福尔波集团的亚麻地板产品目前全世界市场占有率超过 65%。

4.5 Floor material: Forbo Linoneum

To heat the indoor space rapidly, the flooring above the capillary network should be as thin as possible. The material must be hard, wear-resistant, and easy to clean. And it must be visually appealing and easy to install. All those needs can be achieved by Forbo linoneum flooring.

The linen floor has a 150-year history. The earliest form was a "wax" or "oil cloth" that was invented in 1627 and was gradually developed into a floor after more than 200 years. In Scotland in 1860, Walton invented a method for mass production of linseed oil coating and named it Linoleum (in Latin, Linum means linen, while oleum means oil), which can be considered the first linen flooring. Our project with a total of 150 m² uses Forbo Marmoleum for the floor material which is produced and sponsored by Forbo Group (www.forbo.com).

The linen flooring products of the Forbo Group, registered in Zurich with the main production facility in Krommenie, the Netherlands, has worldwide market share of over 65%. The Marmoleum used in this project is only 2.5 mm thick and uses 94%~98% natural materials such as linseed oil, jute, resin, wood flour, limestone, natural mineral pigments and jute backing. Its manufacturing is a purely physical

福尔波集团荷兰克罗默尼生产基地展示中心
Exhibition Center of Forbo Group Production Base, Krommenie, the Netherlands

地坪预制工序的严格执行是亚麻地板质量的保证
The strict process of the floor construction guarantees the quality of flooring surface

本项目使用的梦梵丽亚麻环保地板厚度仅为 2.5mm，使用 94% ~ 98% 的天然材料，如亚麻籽油、黄麻、树脂、木粉、石灰石、天然矿物颜料和黄麻背衬，其生产全部使用可再生能源，是一种纯物理的生产过程，并且实现了从原材料采集、生产到材料使用后废弃的全程绿色环保，为二氧化碳中性产品。由于梦梵丽亚麻环保地板同质透心结构以及天然原材料的特性，它比高分子合成的 PVC 地板和橡胶地板具有更出色的抗老化性能，并且由于具有天然抗氧化的特性，随着时间的推移，表面会越来越耐磨。按中国国标检测，亚麻地板达到难燃材料分级最高标准。

亚麻地板材料还具备一个其他材料所不能匹及的特点，它可以是卷材，长度可达 30 米，拼接缝极细，而且可供选择的

production process that only uses renewable energy. It is a carbon dioxide neutral product.

The linen flooring can be coiled, up to 30 meters in length with countless colors and patterns to choose from. The pieces fit tightly, and the EnergyPLUS Home 4.0 with its clean modern style provides a brilliant showcase for the flooring material. Due to the "box" assembly mode, there is a natural gap. The design adopts two-color splicing for large areas, composed mainly of light gray-green and straight splicing and supported by dark gray-green and arc splicing to maximize the performance of the material.

亚麻地板的现场粘贴　Laying of Forbo Marmoleum flooring on-site

进厅、起居室的地面，整装和曲线拼色效果　The floor pattern in atrium

二层主卧　Master bedroom ,2F

二层走廊　Corridor ,2F

色彩和图案不计其数。正能量房 4.0 简洁明快的现代风格从设计角度为亚麻地板提供了极佳的展示舞台。由于"盒子"的组装模式，拼接处自然形成了缝隙，设计采用大面积双色拼接，以浅灰绿、直线拼接为主，深灰绿、弧线拼接为辅，最大程度发挥材料本身的性能。

亚麻地板从材料供应、预制工厂里的基层加工，到建造现场的表面铺装全部由福尔波集团赞助，并派遣由公司培训的施工队严格遵照产品质量保障技术要求施工完成。

起居室　Livingroom

4.6 卫浴设施

正能量房 4.0 提供的装配式一体化解决方案里采用节水和节约空间的卫浴设备和系统，与全套房屋设备系统一起安装在上下对齐的两个"盒子"里，并且还留出了足够的走道空间以连通左右两侧的盒子。底层和二层的洗手间完全一致，均采用德国德立菲公司赞助提供的隐蔽式水箱系统。这个系统自带钢结构支架，与坐便器侧端相连并支承其重量。墙面层覆盖好之后仅露出冲水面板，而坐便器与地面完全没有接触（悬浮式），方便地面和墙面一体化设计与施工，体现简洁雅致的现代风格。

二层所展示的是开放式洗漱空间和能够欣赏窗外景致的洗浴空间。该设计将通常被安置在卫生空间内的洗漱梳妆空间分离出来。它的私密性较低，将其置于采光效果极好的位置，以增加空间的通融性，并节省人工照明。而且洗漱梳妆空间和洗手间从功能上可被同时使用，提高了空间的功能性。

所有的洁具都是智能及节水的（www. duravit.cn）。

4.6 Sanitary facilities

The water and space saving bathroom equipment is integrated in the prefabricated box system with ample aisle space. The first and second floor washrooms are identical, all using a concealed water tank system sponsored by the German company Duravit. This system comes with a steel frame bracket that connects to the toilet and supports its weight. The toilet is completely out of contact with the floor (suspended), so the flooring is untutched and keeps identical with the flooring in other spaces.

The second floor displays an open wash space and a bathing space with views. The design separates the washable dressing space, which is usually placed in a sanitary space. It offers privacy and excellent natural light to increase the usability of the space and save the use of artificial lighting. Moreover, the washing and dressing space and the restroom can be used simultaneously to improve the functionality of the space. All sanitary devices are smart and water efficient (www.duravit.cn).

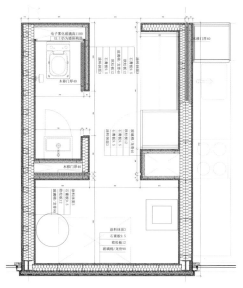

底层卫生间平面　Washroom, 1F

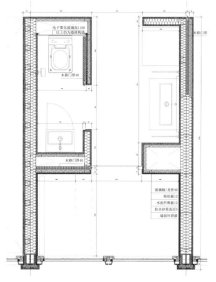

二层卫生间浴室平面　Wash/bathroom, 2F

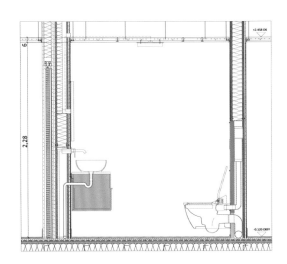

卫生间
W.C.

隐蔽式水箱系统与坐便器　Concealed water tank system and toilet

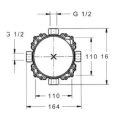

二层开放式洗漱空间
Open wash space on 2F

二层浴室
Bathroom on 2F

KWC
BLUEBOX

4.7 一体化厨房

作为一个家庭的另一个交流场所，在正能量房4.0的理念里，厨房是开放式的。它与餐厅单元连在一起，带有排风设施，出于集中布置管道井空间的考虑，与卫生服务设施的盒子单元毗邻。欧派家居集团股份有限公司（www.oppein.com）的一体化厨房解决方案作为一个独立模块可以完整地嵌入，按照设计订制好的厨房只需要一个下午就安装好了。

4.7 Integral kitchen

The kitchen is open, providing another exchange place for the family. It is connected to the dining unit with exhaust ventilation and is adjacent to the service box, so the pipe well is shared. The integrated kitchen solution from OPPEIN (www.oppein.com) can be completely embedded as a stand-alone module, and the custom-designed kitchen takes only one afternoon to install.

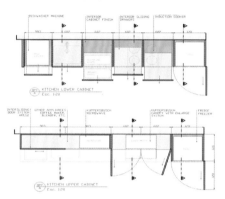

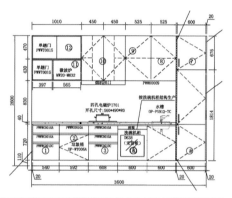

开放式厨房　Open kitchen

4.8 推拉木门

开放式设计理念中的另一个细节是房间门的设计。所有的门均为推拉移门，开启后完全隐藏在墙体内预留的空腔中，以达到"室内无门"的效果（当然也没有门框）。当房间关闭时，门与墙体和地面间的缝隙要尽可能小，以保证隔声效果。这对门的设计和制作都提出了挑战，并且必须保证房屋整体框架结构的精确度。

门的高度与楼层净高相同，铝合金导轨用锚栓固定安装在结构框架梁上，天花板的终饰面层需低于门板的上边缘，这样关起门后上部才看不到滑轮并减小缝隙。最大的门扇是位于底层的厨房门，1226mm×2645mm。由于需要节省墙体内空腔所占空间，门的厚度只有 4.5cm，而且对于板面的平整度要求极高。为保证最佳的隔声和视觉效果，项目经过反复比选，最终采用了上海盼固门窗有限公司加工制作的实木门（全部赞助）。

单扇门的重量达到了 100kg，对铝合金导轨和滑轮钢夹的质量提出了考验。铝合金导轨和滑轮钢夹均由海迪克建筑材料有限公司（www.hideca.com.cn）赞助。

4.8 Sliding timber door

The open concept is reflected by the room door design. All the doors are frameless sliding doors in floor-height and can be completely hidden in the cavity reserved in the wall to achieve the effect of "no door in the room". When a door is closed, the gaps between the door and the wall/floor should be small to ensure sound insulation. The accuracy of the entire frame structure of the house must also be guaranteed.

The aluminum rail is fixed on the structural beam with anchor bolts. The final finish of the ceiling is lower than the upper edge of the door panel, so that there is no gap visible. The kitchen door, 1226mm×2645mm, is the largest one. To save the cavity space inside of the wall, the thickness of the door is only 4.5cm. The solid timber door, manufactured and sponsored by Shanghai Pangu Doors and Windows Co., Ltd., ensures the best sound insulation and visual effects.

The weight of the single door is more than 100kg, which is also a test of the quality of aluminum rails and steel climps, sponsored by Hideca (www.hideca.com.cn).

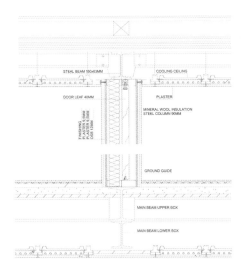

推拉门扇安装节点详图，纵剖面
Details of the door instalation, vertical section

与楼层同高的推拉门扇
Sliding door in floor-height

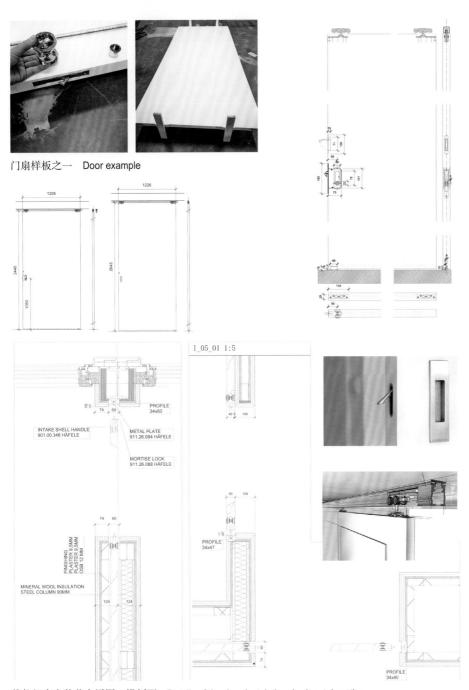

门扇样板之一　Door example

I_05_01 1:5

PROFILE
34x60

INTAKE SHELL HANDLE
901.00.346 HÄFELE

METAL PLATE
911.26.084 HÄFELE

MORTISE LOCK
911.26.088 HÄFELE

PROFILE
34x47

FINISHING
PLASTER 9.5MM
PLASTER 9.5MM
OSB 12 MM

MINERAL WOOL INSULATION
STEEL COLUMN 90MM

PROFILE
34x60

推拉门扇安装节点详图，横剖面　Details of the door instalation, horizontal section

4.9 照明

照明能耗在被动式建筑中占有相当大的比例，正能量房 4.0 虽然通过高效率产能为建筑提供充足的清洁家用电能，但是节约照明用电还是一项能耗指标。首先建筑的通透性为建筑提供了充足的自然采光，白天在室内除了卫生间和设备间外均无需人工照明；在夜间，通过 LED 节能灯为建筑提供充足的照明。

在照明设计中，与擎洲光电科技有限公司（www.letaron-group.com）的设计师们一起工作，项目还整合了健康照明的设计理念，在不同功能空间中采用不同色温光源，来提升人体对室内光环境的健康感受。为了使参观者得到更直观的印象，在起居室的侧墙上，设计并安装了十二个光演示盒子，隐藏在槽口中的灯带发出不同色温的光线。

除了室内照明，室外场地照明也是项目所希望展示的一部分。设计通过沿使用路径布置的条状 LED 灯带发出柔和的室外灯光，用最少的材料和能耗达到最大的使用及艺术效果。项目使用的所有室内外灯具均为擎洲光电科技有限公司赞助。

4.9 Lighting system

Lighting accounts for much of the energy consumption of passive buildings. No artificial lighting is required inside during the day except for the toilets and equipment room. At night, the building is illuminated by sufficient lighting through LED energy-saving lamps.

Working with Letaron Electronic Co., Ltd. (www.letaron-group.com), the project integrates the concept of healthy lighting. It uses light sources with different temperatures in different functional spaces to provide the healthy feeling of the human body. To give visitors a more intuitive impression, 12 light demonstration boxes were designed and installed on the sidewall of the living room. LED-light strips with different light temperatures hidden in the slots emit light of different colors.

In addition to indoor lighting, outdoor venue lighting is part of the project's demonstration. The LED-strips along the visiting paths delivered the functional and artistic effect with minimal material and energy consumption.

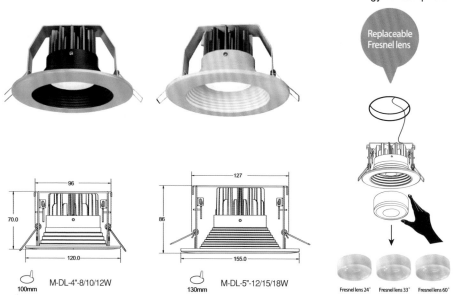

起居室侧墙上的 LED 色温展示盒
LED demonstration boxes on the sidewall in livingroom

建筑夜间照明效果
lighting effect at night

4.10 工程木

正能量房 4.0 虽然在决策时将主体承重结构设定为钢材，但盒子的工作方式对于更为环保的木材也是兼容的。此外，由于需要尽可能避免冷桥，缓冲区与主体结构的连接部位采用工程木将钢焊接连板包在里面的做法，也为工程木从性能和美观角度都提供了一个展示机会（上海至众建筑科技有限公司友情赞助，www.shzzkj.com）。

4.10 Engineered wood

Although the main load-bearing structure is composed of steel, it also works with wood. The buffer zone beams use engineered wood, sponsored by Shanghai Zhizhong Building Technology Ltd. Co.(www.shzzkj.com), to avoid cold bridges caused by the connections with the main steel structure. This provides an opportunity to demonstrate its performance and its aesthetics of engineered wood.

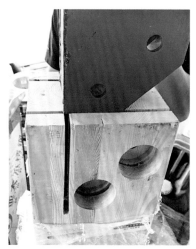

加工好的工程木梁
Engineering beam with cuttings

木结构平台　Timber structure of terrace

4.11 室外场地

生态节能建筑除了建筑体自身以外，还需要充分考虑对周围环境的影响，其中以对原基地生态影响最小为目标。首先，正能量房 4.0 主体建筑采用点式基础，基础建好后原基坑的土壤全部回填，两层盒子结构采用锚栓固定在基础上，高出原室外地坪 30cm。室外木平台也全部架空，采用木桩基，这样建筑对原有基地土壤的影响最小。

其次，在需要做硬地的区位尝试使用科思创透水性地坪涂料解决方案。透水地坪涂料技术是"海绵城市"这一概念的重要组成部分。即使历经户外长期使用，科思创透水性地坪涂料解决方案仍可确保路面的外观和性能，该涂料技术固化速度快，可大大提高施工效率，其高耐候性和高颜色稳定性也可以为地坪带来更多颜色和图案选择，使设计更丰富多彩。

设计中采用黄色和暗红色两种颜色搭配以突出入口部位，车道与人行入口之间留有绿化空隙。但由于面积太小，施工操作不便，最终实施时去掉了小块的暗红色拼接。

4.11 The site

Energy-saving buildings need to minimize impact on the surrounding environment. First, the main building uses a pillar foundation. After the foundation is built, the original soil is all backfilled. The structure of the boxes is mounted on the foundation with anchor bolts, 30 cm higher than the original outdoor ground level. The outdoor wooden platforms are also supported by pillar foundation, so that the construction has the least impact on the original soil.

The project used the water-permeable ground coating solution provided by Covesto in places where hard surface is required. This provides high weather-resistance and color stability.

The design uses yellow and dark red colors to highlight the entrance, leaving a green space between the driveway and the pedestrian access. However, due to the small area, the construction operation was inconvenient, and the dark red stitching of the small pieces was removed in the final implementation.

场地入口透水地坪　Water-permeable ground at the site entrance

技 术
Techniques

$$5$$

正能量房 4.0 的设计理念和被动式建筑不同，它不仅节能，还要尽可能通过建筑和场地生产清洁能量。

The building saves energy and produces energy on its own.

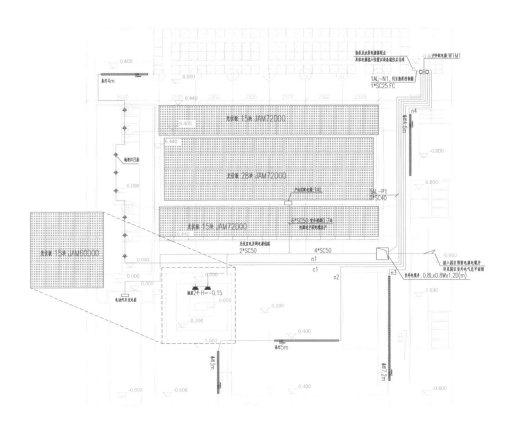

正能量房 4.0 通过建筑和场地生产清洁能量。在阳光充足的时间段，这些能量在满足建筑日常和所有的技术设备使用之需后还有剩余，这些剩余电量被充入电网；在阳光不足的天气，建筑自身所产能量不足以供给其能耗，这时原来存入电网的电量又被取回来用。以年度为计算单位，建筑的产能大于自身能耗。

The building produces energy through its design and construction. During sunny periods, the energy not used by the building will be transmitted to the power grid. On sunless days, the building needs energy. So it will draw electricity from the grid. The annual energy production of the building is greater than its energy consumption.

5.1 制冷

德州地处北半球亚热带圈，属大陆性气候，夏季炎热，最高温度可达近40℃，集中在六、七、八月期间，而且湿度较大，在 80%~90% 之间。夏季要达到室内环境的舒适度，必须采取有效的制冷和除湿措施。

5.1 Cooling

Dezhou is located in the subtropical zone of the northern hemisphere with a continental climate. The hot summers concentrated during June, July and August, can have a maximum temperature of nearly 40 °C, and humidity between 80 and 90%. To achieve the indoor comfort in summer, effective cooling and dehumidification measures are needed.

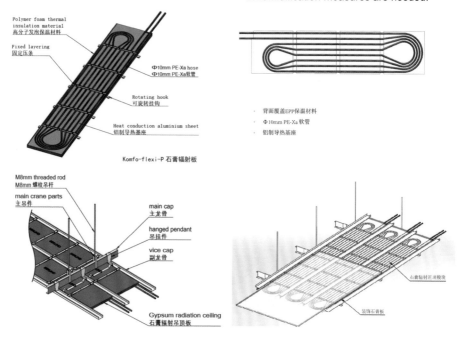

Komfo-flexi-P 系列石膏板结构系统安装图　The installation schame of the Komfo-flexi-P series

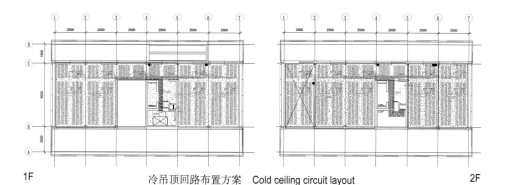

1F　　　　　　　　　冷吊顶回路布置方案　Cold ceiling circuit layout　　　　　　　2F

项目采用冷吊顶制冷方案，与柏瑞德（昆山）环境设备有限公司（http://www.berlind-tech.com）合作制定布置和控制方案。为了适应家居使用的要求，项目使用了石膏板辐射的空调板（Komfo-flexi-P），柏瑞德是项目设计时唯一能够提供此解决方案的单位。

背部带有循环水管道的空调板完全嵌入盒子顶部的结构空间内，这种专门针对居住功能开发的石膏板辐射的空调板在安装完成之后，完全无缝，比VRV空调可节省15.8%用电量。

冷吊顶系统带有电动水阀、温度探头和控制面板，带结露保护。当空气湿度大于结露点时，系统自动关闭从而避免结露。

A ceiling cooling system was adopted for the project, in cooperation with Berlind (Kunshan) Environmental Equipment Co., Ltd. (http://www.berlind-tech.com) Komfo-flexi-P, a gypsum board radiation air conditioning panel, was adapted to home use.

The air-conditioning panel with circulating water pipe on the back is embedded in the ceiling. The gypsum radiation panels are specially developed for residential functions and seamless after finishing work, reducing electricity consumption by 15.8% compared to a VRV air conditioner.

The radiation ceiling system features an electric water valve, temperature sensors and control panels with condensation protection. When the air humidity is greater than the dew point, the system automatically shuts down to avoid condensation.

电动水阀
Electric water valve

温度探头
Temperature sensor

控制面板
Control panel

冷吊顶预制安装现场　Instalation of the rediation panels on the ceiling

5.2 供暖

德州地区不仅夏季炎热，冬季还比较寒冷，一月和十二月的日均最低气温均在0℃以下，极端低温天气可达 –17℃。根据模拟计算，项目采取的被动式措施足以使建筑在冬季无需供暖，但基于保险起见以及项目特殊的展示、试验功能，在设计和实施中都加入了采暖措施。

虽然柏瑞德的冷吊顶系统也可以在冬季转换为供暖模式，通过低温热水从顶部进行热辐射，项目组反复斟酌后还是决定在地板内安装辐射供暖系统。这么做的原因一是地暖舒适度更高，二是可以充分展示盒子系统的能力。对于比德州更为寒冷或是夏季不需要制冷的地区，项目所使用的毛细管辐射热地暖系统都具有参考价值（毛细管网产品由苏州图博节能科技有限公司提供赞助 www.tubesz.com）。

项目在预制时铺设好网垫以及贯通的主循环管道，之后将管道在盒子并接处断开以便盒体运输，在现场再次组装时采用热熔技术再将主管焊接在一起。

5.2 Heating

Dezhou is not only hot in summer but also cold in winter. The average daily minimum temperature in January and December is below 0°C, and low temperatures can reach -17 °C. Simulation indicates that passive measures adopted by the project are sufficient to make heating unnecessary, but a heating system was included to provide for extreme cases and for demonstration and testing of the project.

Although Berlind's cold ceiling system can be converted to the heating mode in winter, the project uses a radiant floor heating system to provide better comfort and demonstrate the capabilities of the box system. For regions colder than Dezhou or where summer cooling is not required, the capillary floor radiant heating system used in the project is a useful reference (sponsored by Suzhou Tubo, www.tubesz.com).

The capillary mats and the main circulation pipes were laid during the prefabrication. After that, the pipes were disconnected at the joints for transport of the boxes. They can be welded together again by reassembling on-site, using hot melt technology.

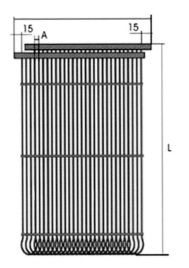

产品型号 Product type: TUBE-CM-III
材料 Materials: 聚丙烯无规共聚物 Polypropylene random copolymer
集水管直径 Water collecting pipe diameter: 20mm×2mm
毛细管直径 Capillary diameter: 4.3mm×0.8mm
毛细管间距 Capillary spacing(A): 10mm
长度 Length(L): 600~8000mm（单位长度 Unit length: 10mm)
宽度 width(B):170~1210mm（单位长度 Unit width: 20mm)

允许热水温度 Allowed hot water temperature: 60°C
工作压力 Operation pressure: 4bar

基层热辐射膜
Thermal radiation film base layer

平铺毛细管网
Laying of capillary mats

与主管连接
Connection to the main pipes

铺自流平
Laying of self-leveling conctrete

5.3 热泵和缓冲水箱

　　冷吊顶和毛细管网地暖里的循环水，最后都进入一个连接热源的缓冲水箱（上海欧特电器有限公司赞助提供，www.ottott.com）。原方案里缓冲水箱可以连接屋顶的太阳能热水装置作为部分热源，基于正能量房4.0的核心理念是将太阳能发电最大化，在深化方案时去掉了太阳能热水部分，精简了设备种类，仅采用一台空气源热泵加热和冷却循环水。夏季热泵的电源全部来自建筑自身利用太阳能发的电。热泵和与之连接的新风除湿机均由上海怡蓝环境科技有限公司赞助并专门为项目度身定做。

　　在缓冲水箱中根据控制系统调整好温度的循环水进入主管道之前，还需要经过各个楼层的分集水器（广东太阳花暖通设备有限公司赞助提供，www.sunfar.cn），它们可以自动或手动控制水温。

5.3 Heat pump and buffer tank

The circulating water, whether in the cold ceiling or the capillary radiation floor, enters a buffer tank and passes the sub-catchment devices, sponsored by SUNFAR (www.sunfar.cn). The tank, sponsored by OTT Electronic Devices (www.ottott.com), is connected to the heat pump. The power of the heat pump comes from the electricity generated by the building using solar energy in summer. This air-to-air heat pump and the dehumidifier are sponsored by Shanghai Yilan Environmental Technology Co., Ltd. and were specially developed for the project.

双冷源除湿机
- 极致新风，主动式新风策略，去除 PM2.5；置换式新风，降低 CO_2 和 VOC 浓度；
- 高效除湿，出风 8g/Kg；
- "恒风量"技术，提升对不同风管的适用性；
- 抽屉式设计，方便维修与安装；
- 专利设计新风和回风阀，比例调节；
- 机组内置冷凝水提升泵
- 极致超薄 –275mm。

辐射用变频风冷冷水机组
- 全变频系统，高效节能；
- 恒定出水温度，针对辐射系统设计；
- 多重降噪技术，机组安静运行；
- 变水温控制，有效避免结露风险。

Double cold source dehumidifier
• Extreme fresh air, active fresh air strategy, remove PM2.5; replaced fresh air, reduce CO_2 and VOC concentration.
• Efficient dehumidification, air output 8g/kg.
• Constant-air-volume technology to improve the applicability serving different air ducts.
• Drawer design for easy maintenance and installation.
• Patented designed fresh air and return air valve, proportional adjustment.
• Built-in condensate lift pump
• Extremely thin - 275mm.

Radiant frequency air-coolling chiller
• Fully variable frequency system, energy efficient.
• Constant outlet temperature, designed for radiation systems.
• Multiple noise reduction technology, quiet operation.
• Variable water temperature control to effectively avoid condensation risks.

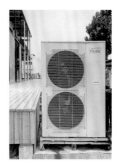

基地东北角的热泵
Heap pump at the NE corner of the site

热泵工作压力表
Pressure gauge of the Heap pump

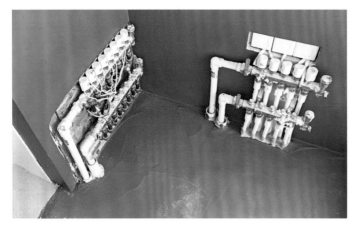

位于二层淋浴室座椅下的分集水器
Sub-catchment located under the seat in the shower room on 2F

底层设备间内的缓冲水箱
The buffer tank in the mechanical room on 1F

5.4 太阳能光伏板

　　太阳能光伏发电是目前发展最为迅速并且前景最被看好的可再生能源产业之一，预计到2050年能够提供全球发电量的11%。从光伏组件的生产情况来看，中国已成为全球主要的太阳能电池和组件制造中心，在全球前20位太阳能电池制造商中，有8家中国大陆企业。为正能量房4.0提供太阳能光伏板支持的合作单位正是其中的佼佼者晶澳太阳能有限公司 (www.jasolar.com)。其双面发电的组件结构，背面发电功率达到正面的65%以上，电力产出比传统的单面板提高3%~15%。项目采用了JAM72D00和JAM60D00两种型号的双面双玻光伏板组件，它们具备出色的低辐照和温度系数表现，以及优异的电力输出能力，从而实现了更低的度电成本。

　　在设计中按照两个方向布置光伏板，全部覆盖透空的缓冲区和实体的屋面。除了必需的检修通道和通风、给排水设备的预留开口外，充分利用了屋面的每一寸面积。实体屋面的面层为东方雨虹提供的白色柔性防水，增强反光度以提高双面光伏板反面的受光率。

　　在二层南北两侧的缓冲区，顶面的双面双玻光伏板不仅能够双面发电，还同时是外围护结构。它能够保证一定的透光率，遮挡风雨的同时在夏季更为缓冲区提供了足够的遮阳防护。

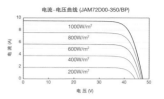

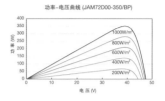

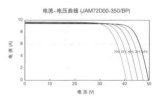

光伏板性能曲线
Performance parameter of JAM72D00
PV-Panels

屋面和缓冲区顶部均采用双面双玻光伏板
The double-sided PV-panels on the roof of the building and buffer zones

5.4 PV-Panels

The photovoltaic panels covering the buffer zones and the solid roof are arranged in two directions. The double-sided PV-panels are from JA Solar Holdings, Co., Ltd., one of the world's leading solar cell and component manufacturers. The panels produce 3%~15% higher power than a traditional single-side panel. The white flexible waterproof provided by Oriental Yuhong increases reflection and the light-receiving rate of the reverse side of the double-sided photovoltaic panels.

In the buffer zones, the double-sided photovoltaic panels on the top not only generate electricity but also are the protection structure. This guarantees light transmittance while providing sufficient sun protection for the buffer zone during the summer.

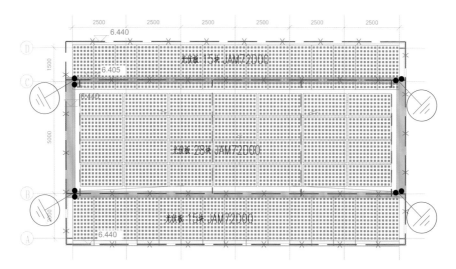

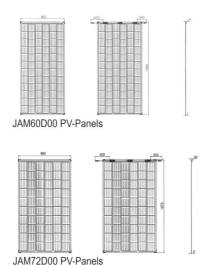

JAM60D00 PV-Panels

JAM72D00 PV-Panels

木凉亭顶部也全部覆盖双面双玻太阳能板
The wooden pavilion covered by double-sided PV

二楼南侧缓冲区双面光伏顶板　Double-side PV-panels of southern buffer zone

预 制
Prefabrication

6

按照项目组织计划，在离比赛建造开始还有三个月的时候，盒子正式开始在工场车间里进行预制加工和预组装。

The prefabrication of the boxes began in the workshop three months before the start of the competition.

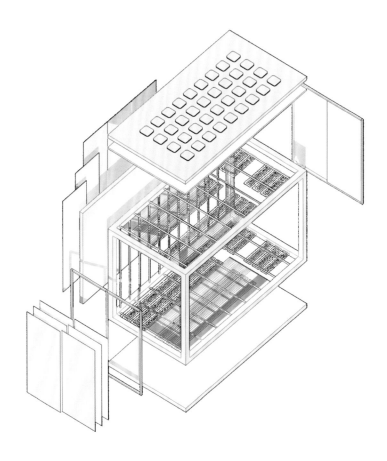

6.1 钢主体承重结构

针对中国住宅市场需求，正能量房4.0 的装配式致力于多层及高层结构体系，单元盒子的承重主体选用重钢结构，柱子与梁均为槽钢 Q235B，柱子断面为 63mm×160mm，底梁为 73mm×200mm，顶梁为 63mm×160mm，采用焊接方式连接。盒子与盒子之间、盒子与基础之间以及顶层盒子与屋面结构之间均采用螺栓连接。

6.1 Steel load-bearing structure

The load-bearing system of the project is based on multi-layer and high-rise structural systems, made of steel. The columns and beams, channel steel Q235B, are welded together. The column section is 63mm×160mm, the bottom beam is 73mm×200mm, and the top beam is 63mm×160mm.

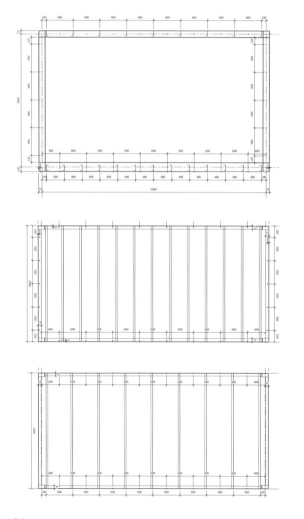

标准盒子结构立面
Side view of standard box

标准盒子顶面图
Top view

标准盒子底面图
Bottom view

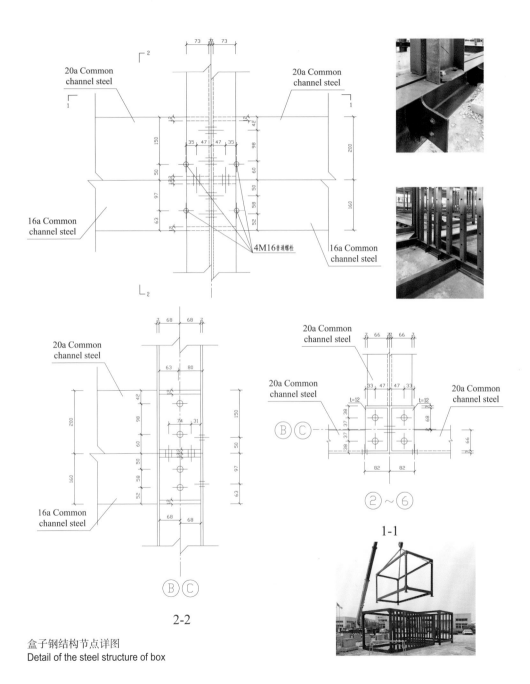

20a Common channel steel

20a Common channel steel

73 73

4M16普通螺栓

16a Common channel steel

16a Common channel steel

20a Common channel steel

16a Common channel steel

2-2

20a Common channel steel

20a Common channel steel

20a Common channel steel

Ⓑ Ⓒ

②~⑥

1-1

盒子钢结构节点详图
Detail of the steel structure of box

6.2 轻钢结构

外墙和非承重内墙、钢承重框架以内的楼地面龙骨均采用轻钢结构，这些构件在预制工场按照设计图纸加工完毕后运至预制组装工地，与主体承重结构通过螺栓固定方式连接。每个盒子都是一个独立的个体，当这些次级承重构件因功能需要进行改动时，可以方便地进行拆卸和重新安装，不会影响到相邻的盒子。所有的钢结构部分均由日照大象房屋建设有限公司（www.anyfunhome.com）承建和赞助，而轻钢结构的设计和制作则是他们特别擅长的领域。

轻钢龙骨结构与 OSB 板基层
Light steel joists & OSB base boards

6.2 Light steel structure

The external walls, non-load-bearing partition walls and the joists in the floor and ceiling are all made of light steel. These components are processed in a prefabricated workshop and connected to the main load-bearing structure with bolts. All steel structures are constructed and sponsored by ANYFUN HOME (www.anyfunhome.com), which specializes in design and production of light steel structures.

盒子主体结构与轻钢龙骨预制组装
Prefabricated assembly of the main structure & the light steel joists of the boxes

6.3 门窗、侧墙

　　钢结构做好后，开始安装在南北两侧外围护构件中比重较大的门窗构件。这些尺寸较大的门窗在工厂预制好后运到预制组装工地，严格按照门窗制造商的安装要求并在他们的指导下被准确地固定在结构框架中。同时东西两侧的实体外墙开始进行基层和玻璃纤维气凝胶毡外保温材料的安装。

6.3 Windows and sidewalls

After the steel structure is completed, the door and windows are installed. These larger doors and windows are prefabricated at the factory and transported to the assembly site. There they are fastened in the structural frame. At the same time, installation begins with the outer layer and the glass fiber aerogel insulation of the solid east and west sidewalls.

侧墙的 OSB 刨花板基层和玻璃纤维气凝胶毡保温层安装
Installation of the OSB base boards and glass fiber aerogel insulation layer of the sidewalls

盒体南侧落地推拉门窗安装
Installation of the sliding windows on the south side of the sidewalls

6.4 顶部暗藏管线、内隔墙

每个单体盒子的顶部 160mm 结构高度内都布有管线，这些可能是风管、照明或弱电线路。在走廊部位是主管线连通处，梁在这个部位根据设计需要开孔洞以便管线穿过。此外，每个盒体的顶部都安装了冷辐射吊顶，可以分开或与邻近的盒体共同控制。

上下水的管道集中布置在设备服务单元处的预留井道中。盒子与盒子之间的内隔墙也装上了 OSB 刨花板基层，它们相互之间没有必然的联系，可以随时根据需要进行互不干扰的改动。

Pipes are placed within the 160 mm height of the top beams in each box. These can be ventilation ducts or used to carry lighting or communication cables. Above the corridor are the main pipelines passing through beams. The radiant ceiling is mounted on the top of each box.

The pipes for water supply and discharge are concentrated in the well in the service box. The interior partition walls are also covered with the OSB boards.

走廊处吊顶内的穿梁主管线
Main pipelines pass through the beams in the ceiling of the corridor

吊顶内的冷辐射板
Cold radiant panels in the ceiling

6.5 底部地暖与面层基层

　　每个单体盒子的底部200mm结构高度内布置了次梁龙骨，之上铺设 OSB 刨花板基层、隔声层（二层盒体）、保温层、毛细管网地暖。之上覆盖钢丝网、自流平水泥。

6.5 Floor heating and floor base layer

The joists on the floor are mounted within the height of 200 mm of the bottom beams of each box. Above are a base layer of OSB board, sound insulation for the second-floor boxes, an insulation layer and capillary floor heating mats. These are covered with steel mesh and self-leveling cement.

右上：地暖保温层
右中：毛细管网之上铺设自流平前的钢筋网
右下：混凝土自流平基层
Upper right:
Insulation of the capillary mats
Middle right:
Laying of the steel meshes above capillary mats
Bottom right:
Concrete self-leveling base

施 工
Construction

7

2018 年 7 月 9 日,预制成半成品的 12 个盒子被拆解开来,用 4 辆 17m 长超高平板车运到比赛场地,使用两台吊车进行吊装。

On July 9 2018, twelve prefabricated boxes with all installations inside were dismantled and transported to the competition site by four 17-meter-long super-high-profile trucks. Two cranes positioned them for reassembly.

正式的比赛建造从 2018 年 7 月 9 日开始至 7 月 31 日结束，此前由组委会负责组织德州当地的施工队伍将基础做好。7 月 9 日，预制成半成品的 12 个盒子被拆解开来，用 4 辆 17m 长超高平板车运到比赛场地，使用租借的两台吊车进行吊装。

The official competition construction period was from July 9 to July 31, 2018. On July 9, twelve prefabricated boxes with all installations inside were dismantled and transported to the competition site by four 17-meter-long super-high-profile trucks. Two cranes positioned them for reassembly.

7.1 组装

按照整体项目的设想，将 12 个在预制工场已经拼装好的盒子重新组装起来是最省力省时的一个环节。但由于基础施工的准确度极低，无论是水平方向还是垂直方向，盒体与基础均无法对位，这给盒体的现场组装造成了极大的困难。在现场施工

7.1 Reassembly

According to the project concept, this process would be very fast. However, due to the low accuracy of the foundation construction, the boxes and the foundation could not be aligned horizontally nor vertically. This caused significant problems.

现浇混凝土基础
Cast-in-place concrete foundation

对基础施工误差的补救处理
Remediation of basic construction errors

基础底板的保温层施工
Insulation construction of the foundation

冒雨卸装、运输预制的盒子
Unload and transport the prefabricated boxes

时不得不采取临时补救措施，如在底层盒子的支撑板上另外打孔、另加钢垫板等。由于每个与基础相连的盒子都要做如此处理，因此底层盒子的现场组装比预计时间大大增加。原来的使用螺栓连接也无法执行，最后采取了将盒子与基础预留钢筋通过焊接方式固定的措施，以保证盒体与基础的连接强度。

从底层西侧开始组装盒子
Assembling from the west side

底层的最后一个盒子
The last box on 1F

从西侧开始安装二层的盒子
Assembly boxes on 2 F from the west side

Temporary remedial measures included additional drilling on the support plate of the bottom boxes, steel backing, etc. Since each box connected to the foundation had to be placed individually, on-site assembly of the underlying boxes took more time than scheduled. Instead of using bolts, the connections were welded to ensure stability.

二层盒子安装的速度很快
The assembly of boxes on 2F was very fast

7.2 缓冲区钢（木）结构安装

　　缓冲区的结构构件由钢立柱、开间之间的联系钢横钢和缓冲区平台的木支撑横梁组成，这些构件都是杆件，由两家不同的单位提供。由于这些构件数目极多、安装耗时，且位于盒子外部，在预制工厂并未预安装。在北侧缓冲区除了平台支承构件，还有一部整体焊接的钢跑梯。这些组件的现场安装需要团队有细致的实施计划、有经验的施工操作和部件的精细制作。

7.2 Steel and wood structure assembly in the buffer zones

The structural members of the buffer zone consist of steel columns, connecting steel beams between the openings, and wooden support beams for the platforms in buffer zones. These components are rods provided by two different producers. Due to the large number of pieces, time-consuming installation, and the fact that they are located outside the box, these components were not installed in the prefabrication workshop. In addition

缓冲区结构安装
Structure installation on the Buffer zone

to supporting beams of the platforms, there is also a welded steel staircase in the northern buffer zone. Still, on-site installation of these components requires a clear implementation plan, experienced construction operations and quality components.

7.3 屋面

　　屋面结构由于需要保温和防水,因此也不能在预制工厂预制。这部分以及在保温和防水做好之后才能安装的屋面太阳能光伏板都需要在现场施工。

　　当屋面防水层做好后,将太阳能光伏板运至屋顶,根据安装位置排放。先进行南北两侧缓冲区顶部的安装,并做好防水密封处理、回路连接;之后进行实体屋顶部分的太阳能板安装。最后进行总线连接及测试。

双面太阳能光伏板运至屋面
The PV-Panels on the roof before installation

双面太阳能光伏板缓冲区安装
The PV-Panels installation on the buffer zones

7.3 Roof

The roof structure was not prefabricated because of the need for waterproofing. Thus the insulation, photovoltaic panels and waterproofing were installed on-site.

When the roof waterproofing layer is complete, the solar photovoltaic panels are lifted to the roof and positioned. The roof of the buffer zones in front of the building are installed first. The gaps between the panels are sealed with the sealants. Finally, the full roof is installed. Then system testing begins.

双面太阳能光伏板屋面安装
The PV-Panels installation on the solid roof

7.4 太阳墙

位于东西侧墙的太阳墙使用空心聚碳酸酯板，由于板材和固定铝框架膨胀系数相差较大，并且板材对气密性的要求较高，这些都对安装质量提出了较高的要求。另外板面较大，现场施工条件有限，也提高了项目的实施的难度。

7.4 Solar walls

The solar sidewalls use hollow polycarbonate sheets. Because of the difference in expansion coefficients between the sheets and their fixing aluminum frames and need to minimize leaks, there is a need for very high quality. Also, the polycarbonate sheets are large, and construction conditions on-site are limited. So this was a challenging process.

7.5 缓冲区推拉扇

缓冲区推拉扇原本是最为快捷的施工部分，但由于对钢结构施工的准确度要求较高，现场施工在这方面的误差较大，再加上施工条件的限制，很难安装位于二层的玻璃推拉扇。好在阳光房在夏季用不到，对建筑在比赛期间的能量表现没有影响。

7.5 Sliding elements in the buffer zones

This was planned to be a fast process. But due to the high accuracy requirements for steel structure construction, the field construction tolerance was too large. On-site construction conditions made it difficult to install the glass sliding elements on the second floor. Fortunately, the solar room is not used in summer, so energy performance of the building during the competition was not affected.

7.6 室内终饰

在现场将所有的技术设施管线连接好后，开始安装石膏板终饰基层、批腻子，最后粉刷。

饰面粉刷干燥后，再按照安装位置开各种照明、通风、控制面板孔洞，安装灯具、风口及各种面板。

7.6 Interior finishing

After technical facilities and pipelines connected on-site, the walls and ceilings are sealed with the gypsum boards and painted.

When the finishes are dry, lamps, air outlets and panels are installed.

7.7 地板面层

在墙面、吊顶大致施工完毕后，开始粘贴亚麻地板面层。之前还要对已经做好的水泥自流平基层再次进行打磨，清理掉其他工序施工时掉落的垃圾和残留物，确保地板的最终效果及粘结牢固度。

7.7 Flooring

After the wall and ceiling are finished, the linen floor finishing is applied. The completed cement self-leveling base layer is polished and cleaned to ensure the appearance and firmness of the floor.

7.8 一体化厨房

厨房的上下水、电、通风管道在预制时留好，当地面面层施工结束后，开始最后一个施工工序，安装运到现场的一体化厨房构件。欧派公司派出一名当地门店的代表负责安装。

7.8 Integral kitchen

After the floor is completed, the kitchen components transported to the site are installed.

7.9 室外平台、凉亭和水池

室外木平台承重结构与建筑主体同时施工，但在盒子和缓冲区钢结构构件吊装完毕后才开始安装面层板。

室外水池的底板和池壁均由当地的施工单位用现浇混凝土预先做好。在现场需要做的是防水和汀步，以及水生植物的栽种。

7.9 Outdoor platform, pavilion, pool

The outdoor wooden platform load-bearing structure is constructed at the same time as the building. Floorboards are installed after placing the boxes and the steel structural members for the buffer zones.

The floor and pool walls of the pool are prepared in advance by a local contractor with cast-in-place concrete. The only on-site work is the waterproofing and the steps. Once the water is filled, aquatic plants are added.

7.10 活动家具

所有的活动家俱均由参赛学生在施工现场用 OSB 板（3cm 厚）手工制作。除了餐桌和衣柜尺度相对较大，其他家具均由边长为 40cm 的两种类型的"盒子"构成：

7.10 Movable furniture

All the movable furniture is hand-made by the participating students on the construction site with OSB boards (3cm thick). Except for the dining table and the large closets, the other furniture consists of two types of boxes

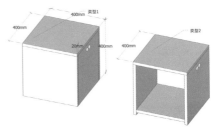

六面均封闭的正方体作为座椅，一个侧面打开的正方体作为矮柜、组合成床（露在外面一侧可以摆放杂物）。这些空心的木头盒子为搬动方便，在两个侧面相对处各钻了两个可以伸入手指的圆孔。

40cm on each side. The seating boxes have the six closed sides. The other boxes, with one side opened, can become beds. Or they can be used as low cabinets or shelves. These hollow wood boxes are easy to move, by putting fingers into the two round holes drilled at two sideboards.

7.11 杂项

当房屋施工完毕，所有的工人都撤离后，剩下分量最大的工作就是清洁：建筑垃圾需要清除，施工时弄脏的玻璃、地板需要擦洗。因为脚手架也拆除了，有些高出部位无法触及，所以未来的用户会需要一部可升降的梯子。

7.11 Miscellaneous

When construction was complete and all the workers left, the biggest job was cleaning: Construction waste needs to be removed, the glass needs to be polished, and the floor needs to be cleaned. Because the scaffolding was gone, some higher areas were inaccessible. Future users will need a ladder.

客厅侧墙上的 LED 展示灯的接线与悬挂
Wiring and hanging of LED display lights on the living room side wall

入口中庭多功能墙上的攀岩点安装
Rock climbing point installation on the atrium multi-purpose wall

实 测
Performance

8

室内湿度必须降到露点以下，冷吊顶制冷机制才会启动，所以必须等建筑的外围护结构施工完毕才能开动热泵，进行系统运行测试。

The radiation ceiling will only start to cool when the indoor humidity is below the dew point. The building envelope must be completed before the heat pump starts the system test.

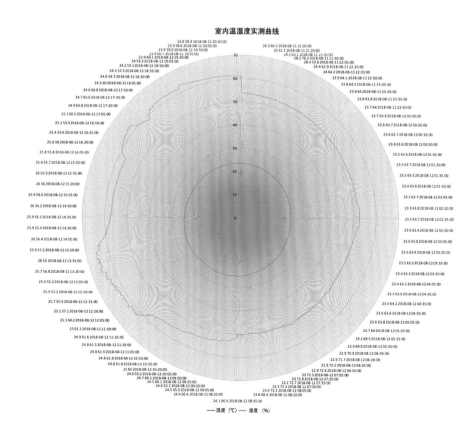

室内温湿度实测曲线

—— 温度（℃）—— 湿度（%）

8.1 室内温湿度

室内湿度必须降到露点以下，冷吊顶制冷机制才会启动，所以必须等建筑的外围护结构施工完毕才能开动热泵，进行系统运行测试。在离规定的建造截止日期还剩 3 天时，所有的设备组件安装且连接完毕，外围护结构也做到了密闭，建筑整体通电后运行调试。

通过反复的运行测试，所有的设备均能正常运行，完成比赛所要求的能耗测试。美中不足的地方是二层的新风不够给力。具体原因主要有两点：一是设计相对紧凑，所留空间余量太少，这也是因为盒子空间本身就很紧凑，必须精打细算；二是管线施工相对粗糙，达不到预制工业化精确度要求，造成较多空间浪费，使得通往二层的风管口径比设计的还要小。这样一来，通往二层的新风量达不到计算量，到末端房间几乎就没有了，二层空间的空气湿度很难仅仅通过系统提供的新风降下来——七月底德州的空气湿度近于饱和，那么冷吊顶系统就不能启动。作为补救措施，二层临时添置了一台小型家用除湿机。

在比赛期间，由于每天的参观人流量较大，每次人流进出建筑都会造成大量空气对流。对由此产生的热损失，系统完全能够承受，而降低因此提高的室内空气湿度则需要新风系统大大增加工作强度。这无疑提高了建筑能耗。好在正能量房 4.0 的主动式产能系统十分给力，德州七八月份的太阳能资源充沛，当然这也是当地气温高、湿度大的主要原因。在整个竞赛期间，建筑大多数时候完全达到了预设的室内气温、湿度、二氧化碳和 PM2.5 数值。对于参观者来说，舒适度体验更是直观的：室内很安静，没有风机盘管的噪声，更没有中央空调风口直接将冷风吹到头上。

8.1 Indoor Temperature and humidity

The radiation ceiling will only start to cool when the indoor humidity drops below the dew point. The building envelope must be completed before the heat pump operates the system test. At the end of the three last construction days, all the parts were installed. The building envelope was sealed, and the building systems could be tested.

All equipment ran normally and met competition requirements. The only problem was that there is inadequate fresh air for the second floor. There were two reasons for this. First, the designed tolerance was too small because of the compact box system. Second, pipeline construction was rough and failed to meet prefabrication requirements. The result was that there is more wasted space, and the duct to the second floor was smaller than the design specified. So humidity on the second floor could not be reduced only by fresh air supplied by the system. At the end of July, humidity in Dezhou was close to the saturation point. So the cold ceiling system could not be started. As a remedy, a dehumidifier was temporarily added on the second floor.

When people went in and out of the building, there was considerable air convection. During the competition, the system could offset heat loss caused by the large number of visitors. But higher indoor humidity meant that the fresh air system had to work harder. This increased the building's energy consumption. Fortunately, the active energy production system of EnergyPLUS Home 4.0 is powerful, and solar resources in Dezhou in July and August were abundant (the main reason for the high temperature and humidity). Throughout the competition, the building nearly always maintained the required indoor air temperature, humidity, carbon dioxide level and concentration of PM2.5. The visitors experienced a comfortable indoor environment: It was quiet, and there was no noise caused by central fan coils. And of course there were no cold drafts from central air conditioning.

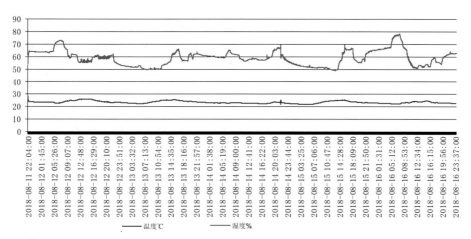

设备信息 Device Info.

设备 Device ID:1

开始时间 Start time: 2018-08-11 21:03:00

温度平均值 Everage Temperature: 23.57

温度最小值 min. Temperature: 21.8

温度最大值 max. Temperature: 28.8

记录条数 Records: 7377

结束时间 End time: 2018-08-16 23:59:00

湿度平均值 Everage Humidify: 58.92

湿度最小值 min. Humidity: 48.9

湿度最大值 max. HUmidity: 78

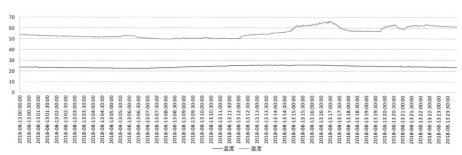

8 月 13 日 00 点 ~24 点室内温湿度实测数据。平均温度 23.9℃，平均湿度 55.29%

Indoor temperature & humidity, 8/13/2018, 00:00~24:00. Everage temperature is 23.9℃, everage humidity is 55.29%

8 月 14 日室外上午建筑阴影处温度
Outdoor temperature at morning 8/14/2018 in shadow

8 月 14 日上午室内实测温度
Indoor temperature at morning 8/14/2018

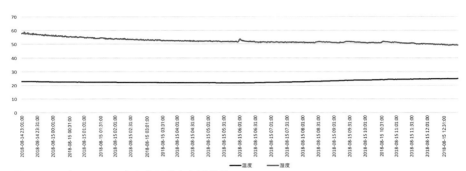

8月14日23点~15日13点室内温湿度实测数据。14个小时期间保持恒温，平均22.8℃。湿度维持在60%以下并缓慢降至50%，平均52.48%

Indoor temperature & humidity, 8/14/2018,23:00~8/15/2018,13:00. Constant temperature. Humidity maintained below 60% and slowly dropped to 50%. Everage temperature is 22.8℃, humidity is 52.48%

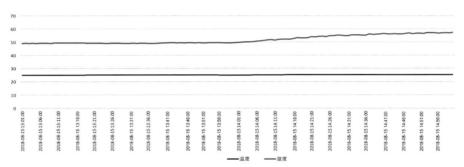

8月15日13点~15点室内温湿度实测数据。正午期间室内温湿度恒定，基本不受外界气温升高影响。14点湿度开始缓慢上升，最高值57.1%，温度恒定在25℃（控制温度）

Indoor temperature & humidity, 8/15/2018,13:00~15:00. Constant temperature and humidity at high noon. Humidity starts to increase slowly to 57.1%. Constant temperature is 25℃

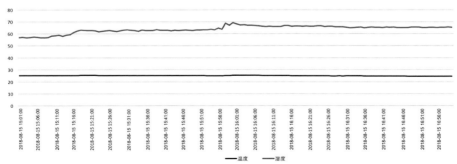

8月15日15点~17点室内温湿度实测数据。14:16室内湿度超过了60%，之后继续缓慢上升，16点达到最大（69.3%）。期间平均湿度也超过了60%，但温度还维持在25℃的设定温度上

Indoor temperature & humidity, 8/15/2018,15:00~17:00. Humidity exceeded 60%, reached max. (69.3%) at 16:00. Everage humidity in the phase also exceeded 60%. Constant temperature is 25℃

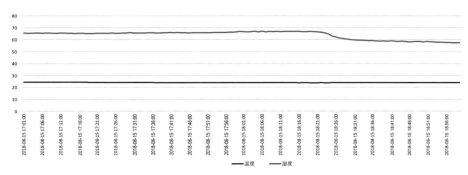

8 月 15 日 17 点 ~19 点室内温湿度实测数据。18:20 室内湿度开始下降，18：30 降至 60% 以下，之后持续下降。温度稍有降低，但基本恒定

Indoor temperature & humidity, 8/15/2018,17:00~19:00. Humidity dropped down from 18:20, reached below 60% at 18:30 and kept to decline. Constant temperature is 25℃

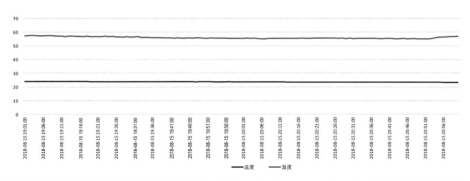

8 月 15 日 19 点 ~21 点室内温湿度实测数据。湿度与温度基本恒定

Indoor temperature & humidity, 8/15/2018,19:00~21:00. Constant Humidity and Temperature

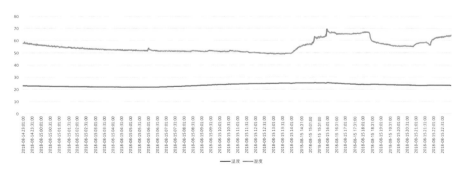

8 月 14 日 23 点 ~8 月 15 日 23 点室内温湿度实测数据。平均温度 23.4℃，平均湿度 55.26%

Indoor temperature & humidity, 8/14/2018,23:00~8/15/2018, 23:00. Everage temperature is 23.4℃, everage humidity is 55.26%

8.2 室内 PM2.5 值

系统采用双冷源除湿机，提供已经去除湿气和粉尘颗粒的新风，且降低 CO_2 和 VOC 浓度。下图为 8 月 13 日 12 点~14 日 12 点室内 PM2.5 浓度实测数据。平均值 0.02455，最大值 0.02688，远低于标准要求，基本维持恒定。

8.2 Indoor PM2.5

Double cold source dehumidifier supplies extreme fresh air, with removed moisture and PM2.5, reduced CO_2 and VOC concentration.

Below is the measured data of indoor PM2.5 concentration, 8/13/2018,12:00~8/14/2018,12:00. Average value is 0.02455, max. value is 0.02688, far lower than required by the rules.

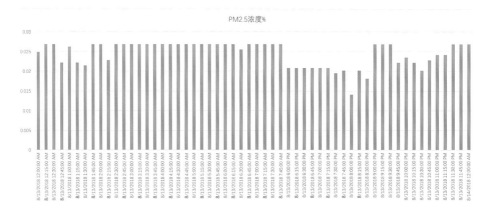

PM2.5浓度%

8.3 产能量和总能耗

为便于管理，在比赛过程中建筑所生产的能量均先输入电网，建筑用电全部取自电网，组委会使用设备自动记录各类电量直到比赛结束。截止到 8 月 16 日晚 10

8.3 Energy production and total energy consumption

To measure the energy efficiency, all electricity produced by the building during the competition was fed into the power grid,

日产能量/日总能耗/能量盈余
Daily energy production/Daily energy consumption/Energy surplus
KWh

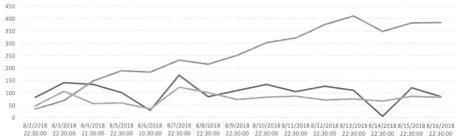

Date	Left meter data		Right meter data	Calculated data				
	E1 [正向有功总] Grid to house	E2 [反向有功总] House to grid	E3 [正向有功总] From PV	E4 [总产能] Total Energy production =E3-Start value	E5[日产能] Daily energy production	E6 [总盈余] Energy surplus =(E2-Start value)-(E1-Start value)	E7 [总消耗] Total nergy comsumption =E4-E6	E8[日总能耗] Daily energy consumption
8/2/2018 22:30:00	117.81	267.47	299.37	80.21	80.21	33.55	46.66	46.66
8/3/2018 22:30:00	179.16	363.61	439.96	220.80	140.59	68.34	152.46	105.80
8/4/2018 22:30:00	218.89	481.41	573.79	354.63	133.83	146.21	208.22	55.76
8/5/2018 22:30:00	253.50	558.29	674.40	455.24	100.61	188.68	266.56	58.34
8/6/2018 10:30:00	274.65	574.10	702.52	483.36	28.12	183.34	300.02	33.46
8/7/2018 22:30:00	344.74	691.91	873.07	653.91	170.55	231.06	422.85	122.83
8/8/2018 22:30:00	407.74	738.10	956.86	737.70	83.79	214.25	523.45	100.60
8/9/2018 22:30:00	446.81	814.00	1065.83	846.67	108.97	251.08	595.59	72.14
8/10/2018 22:30:00	491.47	909.79	1199.32	980.16	133.49	302.21	677.95	82.36
8/11/2018 22:30:00	531.05	967.97	1303.67	1084.51	104.35	320.81	763.70	85.75
8/12/2018 22:30:00	565.83	1057.91	1430.41	1211.25	126.74	375.97	835.28	71.58
8/13/2018 22:30:00	604.33	1130.86	1539.55	1320.39	109.14	410.42	909.97	74.69
8/14/2018 22:30:00	666.63	1130.98	1544.91	1325.75	5.36	348.24	977.51	67.54
8/15/2018 22:30:00	718.52	1216.99	1665.38	1446.22	120.47	382.36	1063.86	86.35
8/16/2018 22:30:00	762.48	1263.49	1750.04	1530.88	84.66	384.90	1145.98	82.12

比赛期间能量平衡数据

点半，建筑在此期间一共发了 1530.88 度电，其中有一天是台风暴雨天气；而建筑的总能量消耗为 1145.98kWh，一共盈余 384.9kWh。

and all electricity that the building consumed was taken from the power grid. The organizing committee used automatic equipment to record these amounts. From August, 2, until 10:30 pm August 16, the building provided 1530.88 kWh to the grid—despite one day of typhoon and heavy rain. Total consumption for the period was 1145.98 kWh. So the building's energy surplus was 384.9 kWh.

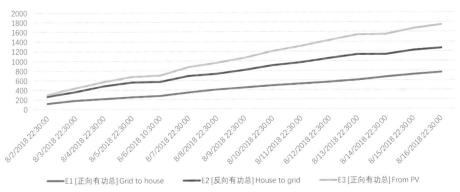

E1 [正向有功总] Grid to house ———— E2 [反向有功总] House to grid ———— E3 [正向有功总] From PV

8月2日~16日能量产出与消耗（单位：千瓦时）　Energy production & consumption Aug. 2~16 in kWh

附录一：同济大学——达姆施塔特工业大学联合赛队成员名单
Appendix 1. Member list of Team Tongji University—Technical University of Darmstadt

项目负责人 Project Leader: 曲翠松 *Cuisong Qu*

指导教师 Teaching Member: 曲翠松 *Cuisong Qu, Anett-Maud Joppien, Christoph Kuhn*, 叶海 *Hai Ye*, 李元齐 *Yuanqi Li*, 郑华海 *Huahai Zheng*, 陈洪斌 *Hongbin Chen*, 潘毅群 *Yiqun Pan*, 李翠 *Cui Li*, 郭海新 *Heinz-Axel Guo*, 范蕊 *Rui Fan*, 邓丰 *Feng Deng*, 张永明 *Yongming Zhang, Susanne Janssen, Edin Saronjic, Emanuel Gießen, Matthias Schönau, Christine Störmer, Frauke Wassum*

参赛学生 Students participant:

队长 Team Leader: 丁蒙成 *Mengcheng Ding*

施工与比赛现场 Construction work & Competition on-site:

建筑学 Architecture: 丁蒙成 *Mengcheng Ding, Sebastian Seibert, Isabella Francesca Baum, Louisa Wenkemann, María Inés El-Hage*, 刘江德 *Jiangde Liu*, 许璐颖 *Luying Xu*, 郭兴达 *Xingda Guo*, 邱融融 *Rongrong Qiu*, 张治宇 *Zhiyu Zhang*, 张效源 *Xiaoyuan Zhang, Michela Vanda Caserini, Alina Theodora Kirsch, Ramona Sabrina Gilde, Nabil Tayyan Torrents*

结构 Structure: 张德峻 *Dejun Zhang*, 梁力 *Li Liang*, 赵珍琲 *Zhenbei Zhao*, 宋清源 *Qingyuan Song*, 王彪 *Biao Wang*

设计参加者 Design work:

建筑学 Architecture: 丁蒙成 *Mengcheng Ding*, 谷兰青 *Lanqing Gu*, 王睿 *Rui Wang*, 朱旭栋 *Xudong Zhu*, 朱傲雪 *Aoxue Zhu*, 戴方睿 *Fangrui Dai*, 程尘悦 *Chenyue Cheng*, 谭炜骏 *Weijun Tan*, 罗淼 *Miao Luo*, 徐婧 *Jing Xu*, 吴卉 *Hui Wu*, 程旭 *Xu Cheng*, 肖宁菲 *Ningfei Xiao*, 韩佩颖 *Peiying Han*, 魏嘉彬 *Jiabin Wei*, 郑兆华 *Zhaohua Zheng*, 周瑾楚 *Jinchu Zhou*, 辛诗奕 *Shiyi Xin*, 罗辛宁 *Xinning Luo*, 李想 *Xiang Li, Claudia Colmo*, 申艺振 *Yejin Shin, María Inés El-Hage, Michela Vanda Caserini, Alina Theodora Kirsch, Ramona Sabrina Gilde, Nabil Tayyan Torrents, Louisa Wenkemann, Sebastian Seibert, Isabella Francesca Baum, Dariya Kryvko, Tizian Naterop, Alejandro Bejarano Tarriño, Carlos Soriano Astorga, Remigio Ortiz Domínguez, Alvaro Garcia de Tiedra Gonzalez, Garcia Adrian Montoro, Maria de la O Molina Perez—Tome, Harun Tören, Tiffany, Chan Hong Tai, Anthony Jendrossek, Gwenaëlle Bihl, Ana Navas Pascual, Angel Cobo Alonso, Oscar Sanchez Rojo, Maria Silva Freire, Simon Groihofer, Stefanie Schramel, Cesar Canadas Fernandez, Clemens Neuber, Matthias Neuber, Nikolaus Punzengruber, Felipe Langa Hernaez*, 邸文博 *Wenbo Di*, 揭喆皓 *Jinghao Jie*, 石文鑫 *Wenxin Shi*, 林辰彻 *Jinchur Lim*, 樊婕 *Jie Fan*, 吴雨 *Yu Wu, José Luis Perez Blanco, Santiago Fajardo Zehni, Nil Palleres Sio, Cornelius Backenstrass, Roman Blum, Annika Bork, Jaroslavna Bruilo, Emel Cakmak, Liyu Cong, Frederik Dauphin, Marina Dimitrova, Mingquan Ding, Matthias Gilles, Andrea Hanak, Orestis Hantjaras, Sarah Herzog, Hee-Yung Im, Kalieb Johanes, Clemens Kauder, Malik Kizilgöz, Niels Langhals, Kasia Mochniak, Artur Ortmann, Manuela Prinz, Philomena Reinmüller, Dominik Reis, Marius Riepe, Katharina Scheurich, Anton Schmunk, Steffen Schwanke, Dominik Schwarz, Andre Seibert, Navid Sharifian, Moritz Trägner, Isabel van Randenborgh, Ronald Volz, Julia Wasik, Jana Weber, Martin Wilfinger, Benjamin Wilkesmann-Altig, Chenqing Xin, Jennifer Yen, Xu Zhang, Stefan Zimmermann*

暖通能量 Mechanics&Energy: 吕岩 *Yan Lyu*

结构 Structure: 朱翌 *Yi Zhu*, 罗睿峰 *Ruifeng Luo*, 郭烽 *Feng Guo*, 黄林 *Lin Huang*, 华凯 *Kai Hua, Chimpanshya Simbeya, Jaume Torres Bonet*

给排水 Plumbing: 周泽婷 *Zeting Zhou*, 邬璐雪 *Luxue Wu*, 江海鑫 *Haixin Jiang*, 贺艺 *Yi He*

暖通电气 Mechanics&Electricity: 郭伟鹏 *Weipeng Guo*, 林伟 *Wei Lin*, 柴佳逸 *Jiayi Chai*, 胡冠伦 *Guanlun Hu*, 顾钦子 *Qinzi Gu*, 陈逸凡 *Yifan Chen*, 康哲 *Zhe Kang*, 尹华琛 *Huachen Yin*, 卓显桂 *Xiangui Zhuo*, 刘锦楠 *Jinnan Liu*, 孙震 *Zhen Sun*, 田辞 *Ci Tian*, 颜哲 *Zhe Yan*, 丁江峰 *Jiangfeng Ding*, 孙魁儒 *Kuiru Sun*, 孟令轩 *Lingxuan Meng*, 何源 *Yuan He*, 孟鸽 *Ge Meng*, 徐璐 *Lu Xu*, 周亿冰 *Yibing Zhou*, 倪诗馨 *Shixin Ni*, 陈祁 *Qi Chen*, 张宇琦 *Yuqi Zhang*

运营 Management: 王浩 *Hao Wang*, 张黎婷 *Liting Zhang*, 周妙明 *Miaoming Zhou*, 郝志伟 *Zhiwei Hao*, 明磊 *Lei Ming*, 钟光淳 *Guangchun Zhong*, 王萍 *Ping Wang*, 蔡东旭 *Dongxu Cai*, 胡雨 *Yu Hu*, 程前 *Qian Cheng,Richie Kihingo Muriuki*, 朱冰洁 *Bingjie Zhu*

北京米兰之窗节能建材有限公司（Milux），保温门窗（Thermal window），www.miluxwindows.com
杜拉维特（中国）洁具有限公司（Duravit），卫浴洁具（Sanitary），www.duravit.cn
日照大象房屋建设有限公司（AnyFun Home），钢结构（Steel structure），www.anyfunhome.com
晶澳太阳能有限公司（JA Solar），光伏组件（PV panel），www.jasolar.com
柏瑞德（昆山）环境设备有限公司（Berlind），冷辐射吊顶（Cooling ceiling），www.berlind-tech.com
上海怡蓝环境科技有限公司（Yilan Tech.），热泵新风机（Heat pump，ventilator）
苏州图博节能科技有限公司（Tube Tech.），毛细管地暖（Capillary floor heating），www.tubesz.com
上海欧特电器有限公司（OTT），缓冲水箱（Buffer water tank），www.ottott.com
广东太阳花暖通设备有限公司（SunFar），分集水器（Water seperator），www.sunfar.cn
东莞市擎洲光电科技有限公司（Letaron），照明灯具（Lighting），www.letaron-group.com
广东希姆乐斯健康照明科技有限公司（Himlux），健康照明（Health lighting），www.himlux.com
福尔波地板（上海）有限公司（Forbo），亚麻地板（Linonium floor），www.forbo.com
阿的克建筑材料有限公司（Hideca），门窗五金配件（Hardware Accessories），www.hideca.com.cn
纳诺科技有限公司（Nano Tech.），气凝胶保温材料（Aerogel thermal insulation），www.nanuo.cn
上海至众建筑科技有限公司（Zhizhong），工程木（Engineering wood），www.shzzkj.com
欧派家居集团股份有限公司（Oppein），整体橱柜（Kitchen），www.oppein.com
上海东方雨虹防水技术有限责任公司（Oriental Yuhong），防水系统（Waterproof system），www.yuhong.com.cn
科思创聚合物（中国）有限公司（Covestro），聚碳酸酯板 & 透水地坪（PCB，water-permeable ground coating），www.covestro.com
上海盼固门窗有限公司（Pangu），木门（Wood door）
同济大学建筑设计研究院（集团）有限公司（Architectural Design），施工费用（Construction cost），http://www.tjad.cn
上海禹颉教育科技有限公司（Yujie Education），BIM 技术咨询（BIM consultancy）

Contest Name 大项名称	Available Points 大项分值	Sub-contest Name 分项名称	Available Points 分项分值	Contest or Sub-contest Type 评审方式
Architecture 建筑设计	100	/	/	J*
Market Appeal 市场潜力	100	/	/	J*
Engineering 工程设计	100	/	/	J*
Communications 宣传推广	100	/	/	J*
Innovation 创新能力	100	/	/	J*
Comfort Zone 舒适程度	100	Temperature 温度	40	M/M**
		Humidity 湿度	20	M/M**
		CO_2 Level 二氧化碳浓度	20	M/M **
		PM2.5 Level PM2.5 浓度	20	M/M**
Appliances 家用电器	100	Refrigerator 冰箱冷藏	10	M/M**
		Freezer 冰箱冷冻	10	M/M**
		Clothes Washer 洗衣	16	M/T***
		Clothes Drying 干衣	32	M/T***
		Dishwasher 洗碗	17	M/T***
		Cooking 烹饪	15	M/T***
Home Life 居家生活	100	Lighting 照明	25	M/T***
		Hot Water 热水	50	M/T***
		Home Electronics 电子设备	10	M/T***
		Dinner Party 晚宴	10	M/T***
		Movie Night 观影	5	M/T***
Commuting 电动通勤	100	/	/	M/T***
Energy Performance 能源绩效	100	Energy Balance 能量平衡	80	M/M**
		Generating Capacity 产能效率	20	M/M**

Note 注: J*=Juried 人工; M/M**=Measured/Monitored 定量监测; M/T***=Measured Task 定量任务

后 记
Acknowledgement

2016 年元旦刚过我接到李振宇院长电话，他问我是否愿意组队参加次年举办的中国太阳能国际竞赛。彼时我正在达姆施塔特工业大学做为期一学期的客座，我当即征询 Kuhn 教授的意见。我们都深知这件事所面临的挑战，尤其是 Joppien 教授，她曾参加过 2005 年在欧洲举办的国际太阳能竞赛，知道其难度。中德联队，面对遥远距离、文化和工作方式胡的区别，一起完成这样的项目更是没有先例。可能是因为面临许多未知的以及完成这项挑战的那种极大刺激，我们最终决定参加。

2018 年夏季全球经历了一个史上罕见的炎热夏天，全球性气候变暖铁证如山。七八月的德州，不是骄阳似火，就是狂风暴雨。和我的学生们一起，我们在德州工地上并肩作战，经历了极端的考验。最后赛队取得了总分第六名的成绩，我们都感到十分欣慰。虽然我们没有拿到冠军，但是我们的建筑坚持使用被动式与主动式相结合的先进节能技术，我们的创新和尝试取得了成功，为应对气候变化做出了我们的努力。

尽管该项目已经完成了一年，但仍然有一些工作没做完，如本书，它记录了参与这个特定项目的所有人的贡献。作为项目的总负责人，我将

此书献给所有为了我们共同的目标而努力过的人。首先要感谢学院领导的信任，将这项重要的任务交付于我；感谢同济大学本科生院以及同济大学建筑设计院提供的资金支持。感谢国内国际二十几家企业提供节能技术与产品、材料赞助，没有他们的全力支持项目将无法得以实施。我们的合作伙伴 Kuhn 教授和 Joppien 教授对于方案的形成做出了极大的贡献，我在学院以及其他相关学院的同事也给予项目组同学很多必要的指导，项目的完成和他们所付出的工作密不可分，郑华海博士和吕岩博士亲自做了结构和暖通能量部分的设计。当然这个项目的成功属于那些参加比赛的学生们，他们是同济大学的中国学生、达姆施塔特工业大学的德国学生，还有来同济大学进行交流的国际生。无论他们参加的是项目的哪个阶段，参加的时间有多长，他们的热情、好学和努力都值得称赞。我尤其要感谢的是丁蒙成，他作为队长把项目从头做到尾，进行了预制过程的现场监督，甚至在他毕业后还坚持完成了比赛，他所付出的工作和辛苦无法衡量。最后要感谢我在夏威夷的美国朋友 Paul，要不是他，本书英文读者的阅读乐趣会大打折扣。

Just after New Year's Day in 2016, I received a request from Prof. Zhenyu Li, Dean of CAUP, if I was willing to participate in the China Solar International Competition to be held the following year. At that time, I was a guest professor at the Technical University Darmstadt. I immediately consulted Professor Kuhn. We all know the challenges of this task, especially Professor Joppien, who participated in the SDC held in Europe in 2005. This time, with a Sino-German United Team, in the face of the geographic distance, differences between cultures and working methods, there was no precedent for such a joint project. Inspired by the unknowns and the challenges, we decided to participate.

The world experienced an extremely hot summer in 2018, which proved that global warming is coming. In July and August, in Dezhou the weather was either burning hot or stormy wet. Together with my students, we fought side by side on the construction site under extreme working conditions. We were pleased that we received the sixth place in the total score. Although we did not win the championship, our building adheres to the advanced energy-saving technology that combines passive and active strategies. Our innovation and experiments succeeded, and we made our own effort to cope with climate change.

Although the project has been finished for one year, there is still work to be done—including this book, which records the contributions of those involved in the project. As the general leader, I dedicate this book to all those who have worked hard for our common goals. I would like to thank the leaders of the college for their trust in giving this important task to me. Thanks to Undergraduate School of Tongji University and Tongji Architectural Design (Group) Co., Ltd. for their financial support. Thanks to more than 20 domestic and international companies for the sponsorship of their energy-saving technologies, products, and materials. Without their full support the project could not have been completed.

Our academic partners, Professor Kuhn and Professor Joppien, made major contributions to the design concept. My colleagues in Tongji University gave students valuable guidance, and the completion of the project would have been impossible without their help. Dr. Zheng Huahai and Dr. Lu Yan personally designed the structure and the HVAC.

Of course, the success of this project belongs to the students who participated in the competition. They are Chinese students of Tongji University, German students of Technical University of Darmstadt, and international students who came to Tongji University for exchange. Regardless of when they joined the project or how long they participated, their enthusiasm, eagerness and hard work are commendable.

I especially want to thank Mengcheng Ding. As the student team leader, he was there from the beginning to the end. He managed on-site supervision of the prefabrication process and insisted on completing the competition after he graduated. His contribution is immeasurable. Finally, I would like to thank my American friend Paul Sturm from Hawaii. Without him, the pleasure of English readers would be considerably diminished.